Armando Gómez
Christian Gutiérrez
Carlos Mendoza

Humans + Machines:

Armando Gómez
Christian Gutiérrez
Carlos Mendoza

Humans + Machines:

the new engineering of work

ScienciaScripts

Imprint
Any brand names and product names mentioned in this book are subject to trademark, brand or patent protection and are trademarks or registered trademarks of their respective holders. The use of brand names, product names, common names, trade names, product descriptions etc. even without a particular marking in this work is in no way to be construed to mean that such names may be regarded as unrestricted in respect of trademark and brand protection legislation and could thus be used by anyone.

Cover image: www.ingimage.com

This book is a translation from the original published under ISBN 978-613-9-46764-8.

Publisher:
Sciencia Scripts
is a trademark of
Dodo Books Indian Ocean Ltd. and OmniScriptum S.R.L publishing group

120 High Road, East Finchley, London, N2 9ED, United Kingdom
Str. Armeneasca 28/1, office 1, Chisinau MD-2012, Republic of Moldova, Europe
Managing Directors: Ieva Konstantinova, Victoria Ursu
info@omniscriptum.com

Printed at: see last page
ISBN: 978-620-8-50043-6

Table of Contents

Welcome to an adventure that explores the present and future of our work, where humans and machines are at the center of an unprecedented transformation. "Humans + Machines: The New Engineering of Work" is not just a book; it is an invitation to discover how technology is redefining who we are and how we work. Throughout these pages, you will find a fascinating journey that begins with the first steps of automation, advances to artificial intelligence and robotics, and goes on to explore the ethical challenges, the changes in skills, and the impact on our well-being.

Each chapter reveals a new piece of this puzzle, showing how our relationship with technology has evolved to make us strategic partners with machines. Imagine a world where creativity and critical thinking are as valuable as programming and data analysis. Here, you will discover the challenges we face and the opportunities this new era offers. This book prepares you to adapt and thrive in a future driven by the synergy between humans and technology, a journey that will transform not only your vision of work, but also of the role we play in this revolution. Prepare to be inspired and open your mind to a world where the limit is only our imagination!

Introduction

The relationship between humans and technology has been a constant in human history, transforming not only our tools, but also our aspirations, values and ways of life. From the first stone tools to complex artificial intelligence systems, each advance has redefined what it means to be human in a technological environment. This documentary book explores this evolving relationship, focusing on how technology has reshaped our work interaction and how human work has changed in step with automation, artificial intelligence (AI), and digitization. In a world where the speed of innovation challenges our abilities to adapt, it is essential to understand how we got here and where we might go in the coming years.

This book examines how our integration with machines began and how it has transformed over time. From the first attempts at automation in manufacturing to the artificial intelligence revolution, this chapter explores how technology has evolved from a mere tool to a collaborative partner in the workplace. With a historical and social lens, this section also questions how technology has altered the perception of human labor from mere operators to true collaborators in a human-machine synergy.

It discusses how automation has redesigned work, eliminating repetitive tasks and optimizing workflows in various industries. This chapter not only examines advances in manufacturing, but also shows the impact of automation tools in the service and finance sectors. In an era where technology is redefining human roles, it also delves into the challenges and barriers organizations face in implementing automation effectively and the effects of these changes on employee well-being.

It also explores how AI is revolutionizing the workplace, particularly in decision-making and complex tasks. With AI, companies can perform more accurate analysis and optimize their organizational performance, giving them a competitive advantage. This chapter presents studies and case studies that demonstrate the impact of AI on productivity and how, by reducing the human burden, it is transforming the concept of efficiency and performance in organizations around the world.

As technology is constantly evolving, Human Adaptation to the Technological Revolution addresses the competencies needed to adapt to this changing environment. In an AI-driven world, skills such as creativity and critical thinking become as essential as technical skills in areas such as programming and data analysis. This chapter discusses the training initiatives needed to prepare the workforce of the future and emphasizes the importance of a continuous learning mindset to stay relevant in an ever-changing job market.

Finally, it analyzes future scenarios and possible business models in a digitized world. Ethics and privacy emerge as crucial issues, especially as companies become increasingly reliant on data and algorithms. Organizational leadership plays a key role in guiding their teams through this technological transition and ensuring that digitization not only optimizes performance, but also maintains the integrity and well-being of employees.

Taken together, this documentary book offers an in-depth, multidimensional exploration of the evolving relationship between humans and technology. Through these chapters, the reader is invited to reflect on the challenges and opportunities that the technological revolution poses for work and society as a whole. By addressing both technical and ethical aspects, this book offers a comprehensive view of today's world of work and a guide on how to navigate a future dominated by synergy between humans and machines.

Chapter 1: The Evolution of the Human-Machine Relationship

Today, it is impossible to imagine a world without technology, without those "extensions" that not only facilitate our tasks, but also expand our capabilities. From the earliest days of humanity, when our ancestors discovered how to use rudimentary tools to survive, to the 21st century, where artificial intelligence and robotics are presented as allies and, in some cases, as competitors in the workplace, the relationship between humans and machines has changed the way we live in extraordinary ways.

To what extent have we let machines integrate into our lives?

The history of this relationship is a narrative filled with inventions, adaptations and breakthroughs, each transforming human life in multiple dimensions. From the first chisels and hammers to the complex machinery of the Industrial Revolutions, technological progress has mirrored the evolution of our societies and priorities. This analysis seeks to delve deeper into this evolution, exploring how each stage has shaped our culture, economy, and the very structure of our daily lives.

In its early stages, technology was something we used strictly as an extension of our own body. The invention of the wheel, the development of agriculture and even fire were advances that gave us control over nature in a previously unthinkable way. These basic achievements gave way to more complex social structures, with hierarchies and roles that increasingly depended on specialized skills and tools. It is fascinating to consider that every small innovation in this prehistoric era paved the way for the great advances that would follow centuries later.

The advent of the Industrial Revolution was a definitive turning point. Machines began to perform jobs that previously required human physical strength, transforming production centers, cities and even the way humans related to each other. Factories not only accelerated production, but redefined work, bringing with them new opportunities as well as challenges. Many feared that machines would replace workers, a concern that, in one way or another, is still relevant today.

Today, the Fourth Industrial Revolution has taken us to a new level of complexity, where machines not only do physical work, but also think and make decisions. In the words of Perez (2020), "the relationship between humans and machines has ceased to be purely instrumental and has become a partnership in which both parties interact and influence each other" (p. 3). This shift has generated endless debates about the ethical limits of artificial intelligence and the place of people in such an automated world.

The impact of this evolution goes far beyond the workplace. Artificial intelligence and robotics are being applied in fields such as medicine, where they facilitate diagnoses and treatments, and in entertainment, generating personalized interactive content. Communications have changed radically, allowing us to be connected to the entire world in a matter of seconds. However, these advances also confront us with new questions: how healthy is this dependence on machines? Are we prepared for the ethical challenges posed by AI?

Throughout this chapter, we explore how the interaction between humans and technology has not always been straightforward or obstacle-free. Martinez (2018) argues that "the digital age and increasing cognitive automation force workers to develop adaptive skills and invest in continuous learning to stay relevant" (p. 50). As technology advances, the need for adaptation becomes even more crucial for those who wish to remain competitive and relevant.

One of the most interesting aspects of this analysis is that technology not only transforms work, but also the way we humans perceive ourselves. With each innovation, what it means to be "human" is redefined and our skills are complemented by the power of machines. The Fourth Industrial Revolution poses a scenario in which human skills such as creativity and critical thinking are valued as much as technical knowledge in an increasingly automated world.

We live in a completely interconnected world. Technology is part of almost all our daily activities, from the smartphones we carry in our pockets to advanced artificial intelligence systems that manage huge databases in real time. Smart homes, virtual assistants and wearable health devices offer us conveniences that, just a few years ago, seemed like something out of a science fiction movie. Technology has ceased to be an external tool and has become an intrinsic part of our daily lives.

In addition, the massive use of technological devices and platforms has opened a space for the creation of new virtual communities. Today, social networks and digital platforms allow people to connect globally, breaking down geographical and cultural barriers. This global connectivity has given rise to social movements, knowledge sharing and mutual support on a scale never seen before. However, it also confronts us with the challenge of discerning between reliable information and misinformation, a problem of increasing relevance in the digital age.

Education is another area where technology has brought about profound changes. Teaching methods have evolved rapidly thanks to digital tools such as online learning platforms, virtual reality simulators and interactive resources that facilitate autonomous learning and access to information from anywhere in the world. This transformation has enabled many people, regardless of their location, to have access to high-quality educational resources, reducing the knowledge gap between different regions.

In the world of work, artificial intelligence and automation have optimized a variety of processes, from manufacturing to executive decision making. AI enables companies to analyze large volumes of data in real time to improve decision accuracy and increase operational efficiency. However, this integration has also raised concerns about job security, as many occupations are expected to be replaced by automated systems. This situation raises important questions about the role of humans in an economy increasingly dominated by technology.

Moreover, privacy and data security have become critical issues as we become more reliant on digital devices that collect and process our personal information. The inclusion of technologies such as facial recognition, geolocation and cloud-based records has raised the challenge of protecting individual privacy in an environment where personal data is a valuable currency. This leads us to consider the extent to which we are willing to sacrifice our privacy in exchange for the conveniences and advantages that technology offers.

Finally, technology has transformed entertainment, expanding our options and providing us with immersive experiences that were previously unthinkable. Augmented reality and virtual reality allow us to live experiences that seem as real as the physical world, immersing us in fictional landscapes or historical simulations. The impact of these advances on our relationships and expectations is profound and, although we are still understanding their long-term effects, it is clear that entertainment will continue to be one of the fields where technology redefines what we consider possible.

The purpose of this chapter is to provide a comprehensive overview of how this relationship has evolved, from the earliest stone tools to sophisticated artificial intelligence technologies. Through a tour of the most important historical milestones and current challenges, we will explore how this relationship between humans and machines has contributed to structural changes that have transformed society. We will also address the impact of the Fourth Industrial Revolution and how it will affect both the labor market and our daily lives.

Analyzing this evolution not only helps us to understand the past, but also to foresee possible paths to the future. If anything has been constant in this history, it is the human ability to adapt. While each advance brings with it both opportunities and challenges, our ability to integrate with technology will be crucial in dealing with the changes to come.

1.1 First steps: from tools to machines

The beginning of the relationship between humans and tools dates back to prehistoric times, when humans began to use rudimentary objects to hunt, gather and build. These tools, although simple, represented a significant advance in the ability of humans to modify their environment and improve their quality of life. From sharpened stones to spears made of wood, the first tools allowed greater precision in everyday tasks and laid the foundations for the development of more advanced technologies. According to Lopez (2019), this was the first great leap in the evolution of human-technology interaction: "Hand tools allowed humans to extend their physical capabilities, which paved the way for the eventual creation of more complex machines" (p. 23).

Over time, these tools evolved in sophistication. During the Bronze Age and Iron Age, humans developed techniques for smelting metals, which allowed for the creation of more durable and efficient tools. This advancement not only changed the way humans worked the land and hunted, but also fostered trade and the expansion of early civilizations. The invention of devices such as the wheel and the lever allowed humans to apply forces more effectively, facilitating transportation and the construction of larger structures. These inventions marked the beginning of a closer, more symbiotic relationship between humans and machines.

The next great technological leap occurred with the emergence of the first industrial machines in the 17th and 18th centuries. Lopez (2019) argues that "the transition from hand tools to machines represented a radical change in the way humans interacted with technology, as machines not only improved productivity, but also began to perform tasks that previously required large amounts of labor" (p. 47). These early machines, though mechanical, relied on human or animal power to function, but they laid the groundwork for future automation.

Among the first machines that revolutionized work were windmills and watermills, used to grind grain and pump water, respectively. These machines enabled ancient societies to generate energy more efficiently, which facilitated agriculture and large-scale food production. Although primitive compared to modern machines, these inventions represented a significant change in the way humans harnessed natural forces to perform everyday tasks.

Finally, towards the end of the 18th century, machines began to incorporate more advanced principles of mechanics and physics, which allowed the creation of more complex and powerful devices. Lopez (2019) highlights that "the invention of the steam engine by James Watt and the subsequent industrial revolution marked a before and after in the history of technology" (p. 62). These early industrial machines not only replaced hand tools, but also transformed the economy and social structure, paving the way for the industrial age.

1.2 The industrial revolution and mechanical automation

The Industrial Revolution, which began in the 18th century in England, was a turning point in human history. It was in this period that machines began to systematically replace human and animal power in production. The introduction of the steam engine, developed by James Watt, was the catalyst for this change. This innovation allowed factories to operate more efficiently and without relying on natural energy sources such as wind or water. Garcia (2020) notes that "the steam engine not only enabled the automation of many industrial processes, but also boosted transportation, which facilitated trade and the expansion of cities" (p. 115).

Mechanical automation brought about a radical transformation in the organization of work. Before the Industrial Revolution, most products were made by hand by craftsmen in small workshops. With the advent of factories and mass production, labor was divided into specialized tasks, allowing for greater efficiency and a reduction in production costs. However, this also had a significant impact on the workforce. Manual labor was replaced by machines, and many artisans were forced to adapt or lose their jobs. Garcia (2020) argues that "the shift toward automation not only altered the economic landscape, but also generated social tensions, as many workers saw their traditional skills become obsolete" (p. 121).

The impact of automation extended beyond factories. Sectors such as agriculture also benefited from the introduction of mechanical machinery. Threshing machines, harvesters and mechanical plows revolutionized agriculture, allowing large-scale food production with less labor. This, in turn, generated a massive exodus of rural workers to the cities, where sought employment in the growing factories. Garcia (2020) mentions that "the Industrial Revolution not only changed the way in which production took place, but also the distribution of the population, marking the beginning of modern urbanization" (p. 128).

However, it was not all positive. Automation also led to longer working hours and deplorable working conditions. The early years of the Industrial Revolution were marked by labor exploitation, particularly of women and children, who worked in dangerous conditions and for miserable wages. This led to the emergence of social movements that fought for better working conditions and for the regulation of child labor. Garcia (2020) points out that "automation brought with it greater productivity, but it also exposed the inequalities inherent in the emerging capitalist system" (p. 134).

As the 19th century progressed, mechanical automation continued to evolve. The introduction of new energy sources, such as electricity, and the development of more sophisticated machines, such as internal combustion engines, allowed automation to expand into new sectors of the economy. Mass production was consolidated as the predominant model, and efficiency became the main objective of engineers and entrepreneurs. Garcia (2020) concludes that "the Industrial Revolution was only the beginning of a process of automation that continues to this day, and that continues to redefine the relationship between humans and machines" (p. 139).

1.3 The digital era: computers and new work relationships

The 20th century brought with it the advent of the digital era, a period characterized by the introduction of computers and information technologies. The development of the first computers, such as the ENIAC in 1946, laid the foundations for a new technological revolution that would profoundly transform the world of work. Computers allowed companies to process large amounts of information quickly and efficiently, which radically changed the way work was organized. Martinez (2018) argues that "with the advent of computers, companies were able to automate administrative and calculation tasks, which allowed for greater efficiency and reduced the need for labor in certain areas" (p. 47).

As computers became more accessible and powerful, they were increasingly integrated into production processes. Factories began to use computerized systems to control machinery and manage inventories, which improved both the accuracy and speed of production. This advance also allowed the creation of new economic sectors, such as software development and information technology. According to Martinez (2018), "the digital era not only transformed production, but also created new employment opportunities in areas that did not exist before, such as programming and systems engineering" (p. 54).

One of the most significant changes brought about by the digital era was the globalization of labor. Information technologies made it possible for companies to operate in different parts of the world simultaneously, which encouraged the creation of global supply chains and the offshoring of production. This, in turn, generated new challenges for workers, who were now competing in a global labor market. Martinez (2018) highlights that "globalization, facilitated by digital technology, has created both opportunities and threats for workers, who must adapt to an increasingly competitive and dynamic environment" (p. 60).

In addition, the digital era brought with it a new way of working: remote work.

Thanks to communication technologies such as email and video conferencing, many employees were able to perform their tasks from anywhere in the world. This not only changed the way companies managed their employees, but also allowed for greater flexibility and work-life balance. However, it also brought new challenges related to information security and digital disconnection. Martinez (2018) explains that "although remote work has improved the quality of life for many employees, it has also generated new forms of stress, such as the difficulty in separating work and personal life" (p. 68).

On the other hand, the digital era marked the beginning of the automation of cognitive tasks. Until then, machines were mainly used to perform physical tasks, but with the development of computers and algorithms, they could now perform tasks that previously required intellectual skills, such as data analysis and decision making. Martinez (2018) states that "cognitive automation represents a new challenge for workers, who must acquire new competencies and skills to stay relevant in the labor market" (p. 74).

In short, the digital era has profoundly changed working relationships, creating new opportunities, but also new challenges. As information technologies continue to advance, the relationship between humans and machines is likely to continue to evolve, further transforming the world of work.

1.4 The fourth industrial revolution: convergence of artificial intelligence and robotics

We are currently immersed in what many experts are calling the Fourth Industrial Revolution, a period marked by the convergence of emerging technologies such as artificial intelligence (AI), advanced robotics, the internet of things (IoT) and 3D printing. These technologies are redefining the relationship between humans and machines in fundamental ways. Perez (2020) suggests that "the Fourth Industrial Revolution is not just about the automation of repetitive tasks, but the integration of intelligent systems that can learn, adapt, and make decisions autonomously" (p. 33).

Artificial intelligence has advanced by leaps and bounds in recent decades, reaching levels of sophistication that allow its application in a wide variety of industries. From automating processes in factories to using algorithms for human resource management, AI is transforming the way companies operate and how employees interact with machines. Perez (2020) highlights that "AI not only increases efficiency, but also enables companies to solve complex problems faster and more effectively, which is changing the dynamics of work in all sectors" (p. 41).

Robotics, meanwhile, has undergone significant advances that have led to the creation of increasingly autonomous and versatile robots. These robots are not only capable of performing complex physical tasks, but can also collaborate with humans in work environments, which has given rise to the concept of "cobots" or collaborative robots. These cobots are designed to work alongside employees, assisting them in tasks that require precision, strength, or repetitiveness. Perez (2020) mentions that "human-robot collaboration is redefining work environments, enabling greater productivity and reducing occupational risks" (p. 48).

One of the most interesting aspects of the fourth industrial revolution is the convergence between AI and robotics, which has resulted in intelligent systems capable of interacting more naturally and efficiently with humans. Advances in natural language processing and image recognition have enabled robots to understand verbal and visual instructions, facilitating their integration into everyday tasks and complex work environments. Perez (2020) argues that "the combination of AI and robotics is creating a new generation of machines that not only perform tasks, but also learn and improve over time, opening up new possibilities for automation and innovation" (p. 52).

However, this technological advancement also poses significant challenges. The increasing automation of tasks through AI and robotics has raised concerns about the future of employment, particularly in sectors that rely on repetitive or manual tasks. Many experts fear that mass automation could lead to the disappearance of millions of jobs, which could exacerbate economic and social inequalities. Perez (2020) warns that "although the Fourth Industrial Revolution offers great opportunities, it also poses significant risks for workers, who must adapt to a constantly evolving labor market" (p. 57).

Finally, the fourth industrial revolution is also generating a change in the skills required in the labor market. As machines take on more complex tasks, workers must focus on developing skills that machines cannot easily replicate, such as creativity, empathy and critical thinking. Employees of the future will need to be prepared to collaborate with intelligent machines, which will require continuous training and constant adaptation to new technologies. Perez (2020) concludes that "in the future, soft skills and the ability to learn continuously will be the keys to thriving in a world of work transformed by AI and robotics" (p. 65).

In this extensive journey through the evolution of the relationship between humans and machines, from the first steps with rudimentary tools to the current convergence with artificial intelligence and robotics, a panorama of unparalleled change, innovation and challenges is envisioned. The history of this interaction has been marked by key moments in which technology has transformed the way we work and live. As technologies continue to advance, the relationship between humans and machines is likely to continue to evolve, posing new challenges and opportunities that will define the future of work.

1.5. Ethics and challenges in collaboration with intelligent machines

The advancement of artificial intelligence (AI) and intelligent machines has raised a number of ethical issues, especially in the workplace and in human relations. These technologies have created new opportunities, but have also raised concerns around privacy, transparency, and accountability. According to Garcia and Rodriguez (2021), the adoption of AI systems in various industries has transformed work roles, as many decisions are delegated to algorithms that operate autonomously. This leads to questions about the human ability to understand and monitor these decisions, especially when applied in complex and ethically sensitive contexts.

A fundamental aspect of collaboration with intelligent machines is the lack of transparency in the algorithms, also known as "black boxes". This opacity makes it difficult to understand the inner workings of AI systems, which generates distrust among users. López and Sánchez (2020) point out that, in sectors such as finance or healthcare, the lack of transparency can have serious consequences, since decisions that affect people's lives are based on criteria that are not always explainable or accessible to those affected.

The autonomy of intelligent machines also raises challenges related to responsibility and accountability. Who should take responsibility in case of errors or damage caused by AI systems? This issue has been widely debated, with authors such as Perez (2019) arguing that liability should lie with both software developers and the organizations that implement these technologies. However, this is not always easy to apply in practice, as autonomous decisions by machines can dilute the chain of liability.

Another relevant ethical issue is privacy and the handling of personal data. AI systems often require large volumes of data to function effectively, which involves constant collection and analysis of personal information. As Fernández and Martínez (2021) point out, this process poses significant risks in terms of privacy, especially when organizations do not implement adequate safeguards to protect user data. In this sense, ethics in data collection and management becomes a crucial aspect to ensure public trust in these technologies.

Algorithmic discrimination is another significant ethical challenge. AI can perpetuate or even amplify human biases present in training data. According to Lopez (2022), when AI systems are trained on historical data that reflect biases or inequalities, they can make decisions that reinforce these patterns, affecting certain groups disproportionately. In the workplace, for example, this can lead to unfair hiring decisions or biases in employee performance evaluation.

Technological dependence and the possible loss of human skills are also points of discussion in this collaboration with intelligent machines. Some experts, such as Gonzalez and Ruiz (2020), warn that automation can lead to the dehumanization of work and the loss of important skills, as people stop performing certain tasks by delegating them to machines. This phenomenon can reduce creativity and problem-solving abilities, essential skills in the digital era.

Another area of concern is the impact of intelligent machines on human autonomy. When key decisions are made by autonomous systems, workers may experience a sense of loss of control over their own functions. According to studies by Ramirez (2018), this can negatively affect employee motivation and engagement, especially in sectors where human interaction is critical. Ethics in human-machine collaboration should include the preservation of individual autonomy as an essential principle.

In terms of security, the implementation of intelligent machines poses risks that are not always easily anticipated. The complexity of AI systems and their ability to learn and adapt can make some behaviors unpredictable. Ortega and Morales (2022) point out that, in critical sectors such as aviation or medicine, it is essential to have protocols that allow humans to intervene and correct errors in emergency situations.

The dilemma between efficiency and ethics is another relevant issue in AI collaboration. While machines can perform tasks faster and more accurately than humans, there are doubts about the appropriateness of delegating all decisions to algorithms, especially when it comes to aspects involving value judgments. As Herrera (2019) mentions, technological efficiency should not always prevail over ethics, and it is important to set limits to ensure that machine decisions respect human values and rights.

Finally, the future of the human-machine relationship requires adequate regulation to address these ethical challenges. Legislators and technology experts must collaborate to create legal frameworks that define the limits of machine autonomy and establish guidelines for transparency and accountability. According to Vega (2020), a clear regulatory framework could help mitigate many of the current ethical risks and contribute to a more equitable and safer collaboration between humans and intelligent machines.

Chapter 2: Automation and the Transformation of Work Processes

Automation is one of the main drivers of modern organizational transformation. As companies adopt advanced technologies to improve efficiency and reduce costs, work processes have undergone profound changes, eliminating repetitive tasks and allowing to focus on higher-value activities. As economists Acemoglu and Restrepo (2020) explain, "automation has made it easier for organizations to optimize their workflows, resulting in greater productivity and accuracy in the execution of routine tasks" (p. 85). This process not only affects manufacturing industries, where automation has been common for decades, but also sectors such as services and finance, where digital technologies have profoundly reshaped their operations.

Automation is not a recent phenomenon. Since the Industrial Revolution, machines have gradually replaced manual labor in many sectors. However, the difference with previous waves of automation is that today's technologies, such as artificial intelligence (AI), machine learning and advanced robotics, can perform cognitive tasks that were previously exclusive to humans. This has generated both excitement and concern, as while some see automation as an opportunity to free workers from tedious tasks, others fear that technological advancement could displace much of the workforce (Bessen, 2019).

The impact of automation on work processes is vast and multifaceted. First, automation has enabled companies to redesign their business processes to maximize efficiency. Instead of relying on human intervention to complete repetitive and error-prone tasks, machines can perform them faster and more consistently. This has led to the creation of more agile and adaptive ways of working, capable of responding to rapid market demands and improving business efficiency (Brynjolfsson & McAfee, 2017).

As companies adopt automated technologies, human roles in the workplace are also being redefined. While automation eliminates some roles, it also creates new opportunities for employees, who must take on more strategic and creative responsibilities. According to Frey and Osborne (2017), "while automation will eliminate some jobs, it will also create new roles that will require unique human skills, such as critical analysis, complex problem solving, and managing automated systems" (p. 258). This change in work dynamics poses significant challenges for training and professional development.

In addition, automation has brought with it the development of specialized technological tools that enable companies to improve their operations. From enterprise resource planning (ERP) systems to collaborative robots (cobots), these technologies are transforming sectors as diverse as manufacturing, financial services and customer service. The adoption of these tools has facilitated the implementation of more agile and efficient business models (Autor & Dorn, 2013).

However, implementing automated technologies is not without its challenges. Companies face multiple barriers, from upfront investment costs to resistance to change from staff. According to Bessen (2019), "organizations wishing to adopt automation must overcome technical, financial, and cultural hurdles to ensure a successful transition" (p. 64). In addition, the integration of new technologies represents a significant challenge, demanding careful planning and change management to minimize disruptions to day-to-day operations.

Another important aspect of automation is its impact on workplace well-being. While automation can reduce physical and mental workload, it can also generate anxiety among employees concerned about the safety of their jobs. According to Brynjolfsson and McAfee (2017), "automation not only affects work in terms of efficiency, but also the psychological well-being of employees, many of whom experience uncertainty about their future in an increasingly automated environment" (p. 193). These concerns raise important questions about how companies can implement automation in a way that benefits both the organization and its employees.

This chapter will provide a detailed exploration of the various aspects of automation and its influence on work processes. First, it will analyze how automation has transformed business processes, eliminating repetitive tasks and optimizing workflows. Next, the new roles and responsibilities of employees in highly automated environments will be explored. Next, the main automation tools in business practice will be addressed, followed by an analysis of the challenges companies face when adopting these technologies. Finally, the impact of automation on workplace well-being will be investigated, assessing both positive and negative aspects.

2.1. Automation and process redesign

Automation has radically changed the way companies design their processes. Instead of relying on manual intervention at every step of a workflow, organizations can now automate repetitive tasks, which not only reduces human error, but also increases the speed and efficiency of operations. According to Davenport and Kirby (2016), "automation allows companies to redesign their workflows to be more efficient and less dependent on human intervention in routine tasks" (p. 45).

One of the greatest benefits of automation is the ability to eliminate repetitive tasks, such as data entry, invoice processing and inventory management. These activities, which traditionally require a great deal of time and resources, can be performed by automated systems that not only execute the tasks faster, but also with greater accuracy. According to Autor (2015), "the automation of administrative tasks has enabled companies to significantly reduce their operating costs and improve the accuracy of their processes" (p. 8).

Automation has also enabled companies to optimize their workflows by implementing systems that manage and coordinate activities more efficiently. Tools such as business process management (BPM) systems allow organizations to model and automate their business processes, which facilitates standardization and continuous improvement. According to van der Aalst (2016), "BPM systems have been instrumental in enabling companies to optimize their processes through automation, eliminating bottlenecks and improving coordination between departments" (p. 27).

In addition, automation has enabled companies to implement more agile and flexible business models. Instead of relying on rigid, linear processes, companies can now react quickly to changes in the market or customer demand by rapidly adjusting their automated processes. According to McAfee and Brynjolfsson (2017), "automation has enabled companies to take a more flexible and agile approach to managing their operations, allowing them to better respond to changing market demands" (p. 115).

A clear example of automation in process redesign is the manufacturing industry. Automated assembly lines, once operated exclusively by humans, are now controlled primarily by industrial robots. Not only can these robots assemble products with much higher accuracy than humans, but they can also work around the clock without a break. According to Bessen (2019), "automation in manufacturing has enabled companies to improve the efficiency of their production lines, resulting in reduced costs and increased productivity" (p. 76).

The service sector has also been profoundly transformed by automation. Robotic process automation (RPA) technologies have enabled financial services firms, for example, to automate repetitive tasks such as transaction verification, account management, and financial reporting. According to Willcocks, Lacity, and Craig (2015), "RPA has enabled service companies to automate a large number of administrative tasks, which has significantly improved the efficiency and accuracy of their operations" (p. 38).

Another sector that has undergone a significant transformation is logistics. The introduction of automated warehouse management systems and autonomous vehicles has enabled companies to optimize their supply chains and reduce transportation costs. According to Wamba et al. (2015), "automation in logistics has improved efficiency in inventory management and transportation, enabling companies to reduce delivery times and improve customer satisfaction" (p. 104).

Automation has also enabled companies to collect and analyze large amounts of data in real time, which has improved decision making. Automated data analytics systems can identify patterns and trends in data, allowing companies to adjust their operations more effectively. According to Davenport and Kirby (2016), "automated data analytics has enabled companies to make more informed, data-driven decisions, which has improved their ability to compete in dynamic markets" (p. 94).

In some industries, automation has led to a complete reorganization of business processes. For example, in the software industry, test automation and continuous deployment have enabled companies to develop and release products faster, without sacrificing quality. According to Bass, Weber, and Zhu (2015), "automation in software development has enabled companies to reduce development times and improve the quality of their products" (p. 58).

In short, automation has transformed business processes, enabling companies to eliminate repetitive tasks, optimize workflows and improve decision making. This transformation has been made possible by advanced technologies such as RPA, BPM systems and automated data analysis, which have enabled organizations to operate in a more agile and efficient manner.

2.2. Human roles in highly automated environments

The increasing adoption of automation has transformed not only processes, but also the roles and responsibilities of employees within organizations. Instead of performing repetitive, low-value tasks, workers today are forced to take on more strategic and creative roles that complement the capabilities of automated machines. According to Frey and Osborne (2017), "automation has displaced many manual tasks, but it has created new roles that require unique human skills, such as critical thinking, innovation, and automated systems management" (p. 263).

One of the major changes in human roles is the need for greater supervision and management of automated machines. Employees are no longer in charge of performing each task manually; instead, they must monitor machine performance and ensure that operations run smoothly. According to Brynjolfsson and McAfee (2017), "workers in highly automated environments must now perform supervisory and maintenance functions, ensuring that machines are running optimally" (p. 203).

In addition, automation has driven demand for technical and digital skills. Employees must be trained to work with automation tools, understand the principles of data analytics, and have a basic understanding of how algorithms and artificial intelligence work. According to Bessen (2019), "as companies adopt more advanced technologies, employees must acquire new technical skills to stay relevant in the job market" (p. 98).

The role of creativity has also become more important in automated environments. Machines can perform repetitive tasks with great efficiency, but they still rely on humans for developing new ideas, innovating products, and solving complex problems. According to Autor (2015), "creativity is a skill that machines cannot yet replicate, making creative employees key assets in automated organizations" (p. 17).

Another important change is the growing importance of social and emotional skills in automated work environments. As machines take over technical tasks, employees must focus on team management, effective communication, and collaboration, skills that are difficult to replicate through automation. According to Deming (2017), "interpersonal and social skills are increasingly valuable in a world of work where technical tasks are being automated" (p. 23).

The rise of automation has also led to the creation of new roles related to the management and development of automated technologies. Automation engineers, data analysts, and artificial intelligence developers are examples of positions that have gained prominence in recent years. According to Brynjolfsson and McAfee (2017), "automation is creating new jobs that require advanced technical skills, such as algorithm development, systems engineering, and big data analytics" (p. 210).

In highly automated environments, employees must also be more adaptable and willing to continuously learn new skills. As technologies change and evolve, employee roles and responsibilities must also adjust. According to Davenport and Kirby (2016), "workers in automated environments must be willing to learn and adapt to new technologies, as job roles will continue to evolve over time" (p. 121).

Another crucial aspect is the collaboration between humans and machines. Instead of completely replacing employees, machines can work alongside them, empowering their capabilities and allowing them to focus on higher-value tasks. According to Bessen (2019), "the real advantage of automation is not to displace workers, but to enable them to collaborate with machines to achieve better results" (p. 132).

Automation has also changed the way employee performance is measured. Instead of evaluating workers based on their ability to complete repetitive tasks, companies now value skills such as problem solving, informed decision making, and the ability to manage complex systems. According to Frey and Osborne (2017), "performance metrics in automated environments have shifted toward cognitive and managerial skills, rather than mere technical ability" (p. 271).

In addition, automation has driven the need for greater cross-functional collaboration. Employees must work closely with technology departments to ensure that automated systems are implemented and managed effectively. According to Brynjolfsson and McAfee (2017), "collaboration between technical and non-technical teams is crucial to the success of automation in organizations" (p. 218).

In short, human roles in highly automated environments have changed significantly. Employees have moved from performing manual tasks to assuming more strategic and creative roles, requiring advanced technical, social, and cognitive skills. This change poses both challenges and opportunities for workers, who must adapt to the new demands of the automated work environment.

2.3. Automation tools in business practice

Automation in the business world has advanced significantly, driven by the development of technological tools that enable companies to optimize their operations. These tools are not only limited to manufacturing, where automation has played a key role for decades, but have also transformed sectors such as services and finance. In each of these sectors, various technologies have been instrumental in achieving greater efficiency and reducing human error, resulting in significant productivity gains.

In the service sector, Robotic Process Automation (RPA) has revolutionized the way companies manage repetitive, rule-based tasks. RPA enables software robots to perform activities such as data entry, document verification and email management. According to Willcocks, Lacity and Craig

(2015) As a result, "RPA has enabled service companies to automate administrative tasks, improving efficiency and reducing processing times" (p. 41). This has been particularly useful in sectors such as customer service, where chatbots and virtual assistants have taken over much of the basic queries, freeing employees to focus on more complex problems.

On the other hand, in manufacturing, industrial robots have been an essential tool in the automation of assembly lines. These robots, capable of working 24 hours a day without a break, have allowed companies to increase production and reduce costs associated with human labor. According to Bessen (2019), "automation in manufacturing has enabled companies to improve efficiency and reduce operating costs" (p. 89). In addition, the introduction of collaborative robots or cobots has facilitated the integration of humans and machines in the same work environment, allowing for greater flexibility in operations.

In finance, automation has been driven by the use of automated trading platforms and artificial intelligence systems for portfolio management and investment decision making. These platforms are capable of analyzing large amounts of data in real time, allowing them to identify investment opportunities and execute trades faster and more efficiently than humans. According to Frey and Osborne (2017), "automation in the financial sector has transformed the way transactions are conducted, improving both the speed and accuracy of trades" (p. 261).

Another key tool in business automation is enterprise resource planning (ERP) software. These systems allow companies to manage their daily operations in an integrated manner, automating processes such as production, logistics, and accounting. According to Davenport and Kirby (2016), "ERP systems have been instrumental in improving operational efficiency by enabling companies to integrate and automate their business processes" (p. 77). This has been particularly useful for large corporations that manage multiple lines of business and need a centralized view of their operations.

In the financial services sector, artificial intelligence (AI) has gained prominence. Not only do AIs enable process automation, but they can also analyze large volumes of data in real time, which helps firms make more accurate predictions about market behavior. According to Brynjolfsson and McAfee (2017), "AI has enabled companies to improve decision making and reduce risk in financial operations" (p. 137).

Logistics automation has also been a growing trend, with the introduction of automated warehouse management systems and autonomous vehicles to optimize the supply chain. These systems not only improve operational efficiency, but also enable companies to reduce delivery times and improve customer satisfaction. According to Wamba et al. (2015), "automation in logistics has enabled companies to manage their inventories more efficiently, resulting in a significant reduction in operational costs" (p. 104).

In the e-commerce arena, automation has played a crucial role in managing inventories and optimizing shipping processes. Automated warehouse management systems allow companies to process orders faster and more accurately, which has resulted in a significant improvement in customer experience. According to Davenport and Kirby (2016), "automation in e-commerce has enabled companies to improve the efficiency of their logistics operations, resulting in increased customer satisfaction" (p. 94).

Data analytics systems are also a key tool in business automation. These systems allow companies to collect and analyze large amounts of data in real time, which helps them make more informed decisions and identify opportunities for improvement in their operational processes. According to van der Aalst (2016), "data analytics automation has enabled companies to make faster decisions based on accurate information" (p. 64).

Finally, marketing automation has enabled companies to improve their relationship with customers by personalizing interactions and optimizing advertising campaigns. Tools such as CRM (Customer Relationship Management) systems make it possible to automate customer relationship management, which improves efficiency and enables greater personalization of services. According to Brynjolfsson and McAfee (2017), "marketing automation has enabled companies to personalize their interactions with customers, resulting in increased loyalty and satisfaction" (p. 115).

In short, automation tools in business practice have transformed the way companies operate, improving efficiency, reducing costs and enabling more agile decision making. From RPA in services to industrial robots in manufacturing to automated trading platforms in finance, automation is redefining the business landscape.

2.4. Challenges in the implementation of automation

Despite the obvious benefits that automation brings to companies, its implementation is not without its challenges. Organizations looking to adopt automated technologies face a number of barriers ranging from high upfront costs to employee resistance to change. Identifying and overcoming these obstacles is critical to ensuring a successful automation process.

One of the main challenges in implementing automation is the initial investment cost. Automated technologies, such as industrial robots or ERP systems, require significant investment in both hardware and software. According to Bessen (2019), "high implementation costs are a major barrier for many companies, especially for small and medium-sized companies that do not have the capital to make such investments" (p. 64). In addition, automation may also involve additional costs related to staff training and integration of new systems with existing processes.

Another major challenge is resistance to change on the part of employees. Many workers fear that automation could jeopardize their jobs, leading to anxiety and resistance to new technologies. According to Brynjolfsson and McAfee (2017), "resistance to change is one of the biggest obstacles to the adoption of automated technologies, as employees fear that their skills will become obsolete" (p. 193). To overcome this challenge, it is crucial for companies to implement training and communication programs that help employees understand the benefits of automation and develop new skills.

In addition, lack of adequate training is another major barrier. The adoption of automated technologies requires employees to acquire new technical skills, which can be a costly and time-consuming process. According to Davenport and Kirby (2016), "the lack of technical skills among employees is a significant barrier to automation implementation, as companies must invest in training programs to ensure that workers can operate and maintain new systems" (p. 121).

Technical complexity can also be a barrier to implementing automated technologies. Companies must ensure that automated systems integrate seamlessly with their existing processes and systems. According to Frey and Osborne (2017), "integrating automated systems can be a complex process, especially in companies that rely on legacy systems that are not compatible with new technologies" (p. 258). This challenge may require upgrading technology infrastructures and hiring automation experts to ensure a successful implementation.

Another major challenge is cybersecurity. As companies adopt automated technologies, they also become more vulnerable to cyberattacks. Automated systems, which often rely on interconnected networks, can be attractive targets for hackers. According to Willcocks, Lacity, and Craig (2015), "cybersecurity is a growing concern for companies implementing automated technologies, as any vulnerability in the systems can be exploited by attackers" (p. 77). To mitigate this risk, companies must invest in robust security measures, such as data encryption and multifactor authentication.

Another challenge related to automation is the disruption of operations during the implementation phase. Companies adopting new automated technologies often experience disruption in their processes while the new systems are being integrated and adjusted. According to McAfee and Brynjolfsson (2017), "transitioning to automated systems can cause temporary disruptions in operations, which can affect productivity and generate additional costs" (p. 103). To minimize disruptions, it is crucial for companies to carefully plan the implementation and conduct thorough testing before going live with automated systems.

Lack of standardization in automation technologies can also be an obstacle. With a wide variety of tools and platforms available on the market, choosing the right technology can be a challenge for companies. According to van der Aalst
(2016) The lack of standardization in automation technologies makes it difficult to select the right tools for the specific needs of each company" (p. 27). Therefore, it is essential that organizations conduct a thorough assessment of their needs before investing in technology.

Another challenge facing companies is the lack of regulatory support. In some sectors, the adoption of automated technologies may be constrained by stringent regulations that have not kept pace with technological advances. According to Frey and Osborne (2017), "outdated regulations can be a major obstacle to automation implementation, especially in sectors such as healthcare and finance" (p. 263). Companies must work collaboratively with regulators to ensure that regulations are adapted to current technological realities.

In addition, the lack of a clear strategic vision can be a major challenge. Many companies rush to adopt automated technologies without a clear strategy, which can result in fragmented implementation and a lack of alignment with business goals. According to Bessen (2019), "automation should be viewed as part of a broader business strategy, rather than a stand-alone solution" (p. 98). To overcome this challenge, companies must develop a clear vision and set specific goals before implementing automated technologies.

Finally, change management is a crucial aspect of automation implementation. Companies must carefully manage the transition process to ensure that employees are engaged and operations are not adversely affected. According to Davenport and Kirby (2016), "effective change management is key to ensuring a smooth transition to automation and minimizing employee resistance" (p. 121).

In conclusion, implementing automated technologies presents a number of challenges for companies, from high upfront costs to employee resistance to change. However, with careful planning, proper training and effective change management, organizations can overcome these obstacles and reap the benefits of automation.

2.5. Automation and its impact on labor welfare

The impact of automation on employee well-being is a topic that has been widely debated in recent years. While automation can improve working conditions by reducing repetitive and hazardous tasks, it can also generate anxiety and stress among employees who fear losing their jobs or seeing their skills obsolete. This subchapter explores both the positive and negative aspects of automation on occupational well-being.

One of the main benefits of automation is the reduction of repetitive tasks that, in many cases, are monotonous and exhausting both physically and mentally. According to Brynjolfsson and McAfee (2017), "automation has freed employees from routine tasks, allowing them to focus on more creative and rewarding activities" (p. 137).

This not only improves productivity, but can also increase job satisfaction, as employees feel they are making a more meaningful contribution to the organization.

In addition, automation can improve occupational safety by eliminating dangerous tasks that previously had to be performed by employees. In sectors such as manufacturing and mining, robots can take on jobs that involve physical hazards, reducing the incidence of workplace accidents. According to Frey and Osborne (2017), "automation in hazardous environments has significantly reduced the number of workplace accidents, thereby improving employee well-being" (p. 263).

Another positive aspect of automation is labor flexibility. With the implementation of automated technologies, many employees have the ability to work remotely or on more flexible schedules, allowing them to better balance their work and personal responsibilities. According to Davenport and Kirby (2016), "automation has enabled companies to offer more flexible work options, which has improved employees' quality of life" (p. 94). This flexibility can be particularly beneficial for employees who have family responsibilities or who prefer to work in a nontraditional environment.

However, automation can also have negative effects on employee well-being. One of the biggest problems is anxiety about the future of work. As machines and algorithms take over more tasks, many employees fear that their jobs may be replaced by technology. According to Bessen (2019), "uncertainty about job security is a major concern among employees in highly automated environments" (p. 64). This anxiety can negatively affect workers' mental health, which in turn can reduce their motivation and productivity.

In addition, automation can generate a sense of alienation among employees, especially those who see their tasks reduced to machine supervision. According to Frey and Osborne (2017), "automation can make employees feel disconnected from their work, as tasks they once actively performed are now managed by machines" (p. 258). This alienation can negatively affect employee morale and commitment to the organization.

Another negative aspect is the stress associated with adapting to new technologies. As companies adopt automated technologies, employees must learn new technical skills to operate and maintain the systems. According to Brynjolfsson and McAfee

(2017) The need to acquire new skills can generate stress among employees, especially those who are unfamiliar with technology" (p. 193). This stress can be particularly acute among older employees, who may feel less able to adapt to the new demands of the job.

Lack of control over work is another problem that can arise as companies automate their operations. In some cases, employees feel they have lost control over their work as important decisions are made by algorithms or automated systems. According to Davenport and Kirby (2016), "automation can make employees feel that they have less autonomy in their work, which can negatively affect their emotional well-being" (p. 121).

Despite these challenges, companies can take steps to mitigate the negative effects of automation on employee well-being. One of the most effective strategies is ongoing training. According to Willcocks, Lacity, and Craig (2015), "training in new skills is critical to help employees adapt to automated environments and reduce anxiety about job security" (p. 77). By offering training programs, companies can empower employees to feel more confident in their roles and more prepared to deal with technological changes.

Another important strategy is open and transparent communication. According to McAfee and Brynjolfsson (2017), "companies adopting automated technologies should maintain open communication with their employees to address their concerns and ensure they understand the benefits of automation" (p. 203). This can help reduce anxiety and foster a more positive organizational culture.

In addition, companies must involve employees in the automation process. According to Brynjolfsson and McAfee (2017), "when employees are involved in the implementation of automated technologies, they are more likely to feel engaged and valued by the organization" (p. 218). Not only does this improve employee well-being, but it can also contribute to the success of automation, as employees can provide valuable feedback on how to optimize processes.Chapter Conclusion.

Automation has profoundly transformed the service, manufacturing and finance sectors, driving significant improvements in efficiency, accuracy and safety. However, its implementation presents diverse challenges including resistance to change, technical issues and regulatory challenges. As companies move forward with the adoption of these technologies, it is critical that they implement strategies to minimize their negative impact on workplace well-being by adapting their practices to maximize the value of automated tools and ensure a positive work environment.

Chapter 3: Artificial intelligence and its impact on work engineering.

Artificial intelligence (AI) is currently considered one of the most disruptive technologies in the field of work engineering, transforming the management, optimization and execution of industrial and organizational processes in an unprecedented way. This technological revolution, characterized by the development of deep learning algorithms and machines that can learn from experience, has opened up new possibilities in automation, predictive analytics and real-time decision making. Work engineering, a field traditionally focused on improving efficiency and productivity through process optimization and resource management, has found in AI an ally to carry out tasks with an unprecedented level of precision, speed and adaptability (Brynjolfsson and McAfee, 2014).

Traditionally, work engineering has been based on techniques such as time and motion analysis, workstation design and resource optimization. However, with the advancement of AI, many of these methodologies have been supplemented or replaced by technologies that enable more comprehensive, real-time analysis of large volumes of data. Today, AI enables engineers to develop advanced systems that optimize performance, anticipate failures, suggest improvements, and adjust to changing conditions autonomously. In this sense, AI significantly expands the scope and impact of engineering work, transforming the static optimization approach into a more dynamic and adaptive one (Daugherty Wilson, 2018).

The impact of AI on work engineering can be observed in various application areas. In manufacturing, for example, artificial intelligence facilitates advanced process automation and enables more efficient integration of production systems. Through the implementation of sensors and the use of machine learning algorithms, engineers can monitor operations in real time, foresee problems before they occur, and make automatic adjustments to optimize machine performance. This not only improves efficiency, but also reduces costs and minimizes the risk of costly system failures (Russell and Norvig, 2021).

Another important area of impact is business decision making. AI has changed the way data is analyzed and applied in business decisions, achieving more agile and accurate decision processes based on large volumes of data and reliable predictions.

Through deep learning algorithms, AI systems can analyze complex patterns in large volumes of data and provide recommendations based on reliable predictions. This is particularly useful in job engineering, where decisions about resource allocation, production scheduling and supply chain management require in-depth analysis of multiple variables. Thanks to AI, these decisions can be made in a more informed and strategic manner, leading to overall resource optimization and improved organizational competitiveness (Frey and Osborne, 2017).

In addition, AI also acts as an assistant in complex tasks that require precision, such as system design, fault detection and quality monitoring. Rather than relying solely on the engineer's knowledge and experience, AI systems can assist in performing detailed technical analysis, detecting system anomalies and providing real-time solutions. This support allows engineers to focus on higher-value tasks, such as planning and innovation, leaving AI to manage routine or precision-intensive operations. In this way, AI functions not only as a support tool, but also as a partial substitute in areas of high specialization, improving efficiency and reducing margins of error (Autor et al., 2003).

Finally, the impact of AI on productivity and organizational performance has become evident in multiple studies and success stories showing how AI implementation contributes to business efficiency and competitiveness. AI's ability to analyze large amounts of data and make adjustments in real time allows organizations to respond faster to changes in the environment and take advantage of optimization opportunities on an ongoing basis. Improvements in organizational performance are particularly significant in process-intensive sectors, such as manufacturing, logistics, and healthcare, where the use of AI can reduce operating costs, increase production capacity, and improve the quality of products and services (Acemoglu & Restrepo, 2018).

However, despite the obvious benefits of AI in work engineering, this technology also presents significant challenges. One of the biggest challenges is the integration of AI into complex organizational systems, which often require advanced digital infrastructure and trained personnel to monitor and adjust AI systems. In addition, AI implementation raises ethical questions related to data privacy, information security and the impact on human employment. AI-driven automation has the potential to reduce demand for certain traditional roles, which poses challenges in terms of workforce adaptation and training.

As AI continues to evolve and expand its reach, it will be critical for organizations and engineers to find a balance between technological innovation and developing a resilient and adaptable workforce (Brynjolfsson and McAfee, 2014).

Artificial intelligence represents a paradigm shift in work engineering, offering a series of tools and capabilities to optimize processes, improve decision-making and increase productivity. However, to take full advantage of these benefits, organizations must face the challenges of integrating this technology and adopt responsible and strategic approaches to its implementation. This chapter will explore the practical applications of AI in work engineering, examining its role in decision making, its use in complex tasks, and its impact on organizational performance. In doing so, it aims to provide a comprehensive view of how artificial intelligence is reshaping the field of work engineering and the challenges that remain to be addressed.

3.1 AI applications in occupational engineering

In the field of work engineering, artificial intelligence has revolutionized resource optimization, process improvement and accurate and efficient decision making. AI applications in this area range from supply chain management and predictive maintenance of equipment to real-time process optimization and automation of repetitive tasks. These technologies have helped redefine the boundaries of productivity and efficiency, offering adaptive solutions that enable companies and engineers to improve organizational performance and reduce operating costs (Daugherty & Wilson, 2018). The following examines some of the key areas where AI is having a considerable impact on work engineering.

One of the main objectives of work engineering is to optimize industrial processes to achieve maximum performance with the available resources. In this regard, AI has played a crucial role in enabling companies to analyze large volumes of data in real time and use that data to adjust and improve their operations. Through machine learning algorithms, companies can identify patterns and trends in their production processes that would not be evident with traditional methods. This allows for precise adjustments in areas such as energy consumption, production scheduling, and resource allocation (Russell & Norvig, 2021).

For example, in manufacturing, AI makes it possible to analyze data in real time, identifying bottlenecks and anticipating failures, thereby minimizing downtime and improving operational efficiency. In this way, companies can reduce downtime and increase the efficiency of their operations. A case study in the automotive industry shows how one manufacturer used AI to analyze its assembly processes and reduce production times by 15% by optimizing workflow and minimizing waiting times between stations (Brynjolfsson & McAfee, 2014). This type of application is especially useful in environments where demand is variable and rapid adaptation is required to avoid waste and maximize resource utilization.

Predictive maintenance is one of the most relevant AI applications in work engineering, as it makes it possible to anticipate equipment failures and take preventive measures to avoid production interruptions. Through sensors installed on machinery and equipment, AI systems can collect real-time data on equipment performance and condition. This data is analyzed by machine learning algorithms that identify behavioral patterns and alert operators to potential problems before they occur. This technology allows companies to reduce costs associated with reactive maintenance and avoid productivity losses due to unplanned downtime (Frey & Osborne, 2017).

One prominent example is the use of predictive maintenance in the aviation industry, where AI systems continuously monitor the condition of engines and other critical aircraft components. These systems can detect small anomalies in engine performance, allowing airlines to schedule maintenance before a failure causes delays or poses a safety risk. This approach not only improves safety and reliability, but also reduces operating costs and allows airlines to operate more efficiently (Acemoglu & Restrepo, 2018).

Automation of repetitive tasks is another area where AI has proven particularly useful in work engineering. Through the use of AI systems and collaborative robots (cobots), companies can automate repetitive and monotonous tasks, freeing up human workers to focus on higher value-added activities. This automation not only increases efficiency, but also improves job satisfaction by reducing the physical and mental workload of employees (Daugherty & Wilson, 2018).

In logistics and warehousing, AI is integrated with robots to perform tasks such as material movement and inventory management, increasing accuracy and reducing processing times in each operation. Amazon, for example, has implemented robots in its distribution centers to reduce the time workers spend moving goods from one location to another. These robots operate autonomously and collaborate with human employees to complete tasks faster and more efficiently, allowing the company to reduce processing times and improve customer satisfaction (Frey & Osborne, 2017).

Artificial intelligence has also revolutionized supply chain management, an essential function in work engineering that involves coordinating the production, storage and distribution of products. AI systems enable companies to analyze data from across the supply chain and optimize logistics to reduce delivery times and minimize costs. Through the use of machine learning algorithms, companies can forecast demand, plan inventory, and manage distribution routes more efficiently (Russell & Norvig, 2021).

For example, in the retail industry, AI systems can predict product demand based on historical sales patterns, seasonal factors and other relevant data. This allows companies to adjust their inventory and avoid both excess and shortages of products, thereby optimizing storage costs and increasing product availability for the consumer. Companies such as Walmart and Target use AI to manage their supply chains efficiently, reducing logistics costs and improving customer satisfaction by ensuring product availability in their stores and online platforms (Brynjolfsson & McAfee, 2014).

Workplace safety is a central concern in occupational engineering, and AI has proven to be an effective tool for improving safety standards and reducing the risk of accidents. Through real-time data collection and analysis, AI systems can identify hazardous conditions in the work environment and alert workers and supervisors before an accident occurs. In addition, computer vision systems can monitor work areas and detect dangerous behaviors or violations of safety regulations, sending immediate alerts to correct the problem (Daugherty & Wilson, 2018).

One example is the use of AI in the construction industry, where computer vision systems analyze images and videos in real time to detect unsafe conditions, such as lack of personal protective equipment or the presence of hazardous materials. These systems can alert safety supervisors and reduce the risk of accidents by ensuring that workers comply with safety protocols. In industrial settings, AI systems can also foresee fire hazards or structural failures, allowing companies to take preventative measures to protect both employees and assets (Frey & Osborne, 2017).

Workforce planning and management is another area where AI has proven to be effective in work engineering. By analyzing data on employee performance, availability and workload, AI systems can help companies optimize their schedules and more evenly distribute workloads. This allows for reduced burnout and improved worker morale, while ensuring that the company is adequately staffed for each shift (Acemoglu & Restrepo, 2018).

An example of this application is in the hospital sector, where AI is used to optimize shift scheduling for doctors and nursing staff, ensuring that there is enough staff to meet the needs of patients at all times. This type of system reduces stress on healthcare staff and improves the quality of patient care by ensuring that the hospital is adequately equipped to handle demand (Brynjolfsson & McAfee, 2014).

In quality control, AI dramatically improves product inspection accuracy, surpassing human ability in defect detection and ensuring high quality standards in final production. Through computer vision systems and deep learning algorithms, companies can identify product defects at a higher speed and accuracy than human inspectors. This not only improves product quality, but also allows companies to reduce the time and costs associated with manual inspections (Russell & Norvig, 2021).

In the electronics industry, for example, AI systems inspect circuit boards for microscopic flaws that might go unnoticed by human eyes. These systems are able to detect defects in real time, allowing companies to correct problems before products reach the market. This type of automated quality control helps companies maintain high standards and reduce the costs of returning or repairing defective products (Daugherty & Wilson, 2018).

The applications of artificial intelligence in work engineering span a broad spectrum of functions and processes, from optimizing production to improving occupational safety. These technologies have enabled engineers at work to overcome the limitations of traditional methods of analysis and optimization, providing advanced tools to manage resources, anticipate problems, and improve efficiency. Although there are challenges in implementing AI, especially in terms of cost and integration with existing systems, the benefits in terms of productivity, cost reduction and quality improvement justify its adoption in most industries.

AI not only transforms the way tasks are performed, but also redefines the role of work engineers, who must now learn to work alongside these technologies and adapt to rapid advances in the field. As AI continues to evolve, its influence on work engineering is likely to continue to grow, opening up new opportunities to improve efficiency and sustainability in industrial and organizational settings.

3.2 AI and business decision making.

Artificial intelligence (AI) has revolutionized business decision making, enabling organizations to analyze large volumes of data, derive hidden patterns and optimize decision making in real time. In work engineering, AI plays a key role in improving operational efficiency, reducing costs and optimizing resources, facilitating more informed and strategic decision making. The implementation of AI systems has not only transformed decisions at the operational level, but also at the strategic level, helping leaders to formulate business plans and policies based on accurate data and the predictive capabilities of these systems. In this context, AI has become an indispensable tool that redefines the way organizations address their challenges and opportunities.

AI and real-time data analysis

One of the main contributions of AI to business decision making in work engineering is the ability to analyze data in real time. AI systems make it possible to collect, process and analyze data as it is generated, making it easier for companies to respond to changes in the environment immediately. In sectors such as manufacturing, for example, AI enables real-time monitoring of machine conditions, detecting anomalies and anticipating potential failures before they occur. By identifying patterns in machine behavior, AI systems can recommend immediate adjustments to avoid production losses and reduce maintenance costs (Brynjolfsson & McAfee, 2014).

Real-time analytics capabilities not only help in solving operational problems, but also improve overall process efficiency. For example, in a production plant, AI systems can constantly monitor inventory levels, production line efficiency and final product quality. If a problem is detected, the AI system can alert supervisors and suggest adjustments to the production process, thus optimizing performance without the need for constant human intervention. In this way, AI becomes a key ally in work engineering, helping engineers make informed decisions that improve productivity and product quality (Frey & Osborne, 2017).

Resource optimization and data-driven decision making

Resource optimization is another area where AI has transformed decision making in work engineering. AI systems enable companies to analyze the use of resources such as time, personnel and production materials, and find opportunities for improvement in the distribution and allocation of these resources. For example, machine learning algorithms can analyze production patterns and suggest adjustments to task scheduling to reduce downtime and maximize the use of available resources.

A notable case of this application is in supply chain planning. AI can analyze real-time and historical data to forecast product demand and adjust production and inventory accordingly. This enables companies to reduce warehousing costs and avoid shortages or excess inventory, thus optimizing supply chain efficiency. In sectors where margins of error are narrow and competition is high, such as in the industry retail or logistics, AI provides a competitive advantage by improving forecast accuracy and reducing the risk of economic losses due to inefficient resource management (Russell & Norvig, 2021).

In addition, AI enables advanced simulations that help engineers anticipate the impact of different decisions on organizational performance. By using predictive models, companies can test different scenarios and evaluate which one will generate the best results before implementing a decision. This type of simulation is particularly useful in large-scale projects, where an error in planning can result in significant losses. The ability to predict outcomes before implementation allows decision makers to reduce risks and optimize the use of resources, making decisions with a high degree of certainty.

AI in risk management and scenario assessment

Risk management is an essential component of decision making in engineering work, especially in highly complex environments with limited margins for error. AI facilitates risk assessment by analyzing historical and current data to identify patterns and trends that may indicate potential risk. Through deep learning algorithms, AI systems can foresee incidents or problems before they occur, allowing decision makers to act preemptively and minimize the impact of risks (Daugherty & Wilson, 2018).

In the construction industry, for example, AI systems can analyze data related to weather, material availability, and machinery performance to anticipate problems that could delay the project or increase costs. If the AI system detects conditions that could indicate imminent risk, it can recommend corrective actions, such as adjusting the schedule or increasing safety measures. In this way, AI not only improves decision-making by providing accurate and timely data, but also enables more effective risk management and proactive risk management.

AI also facilitates scenario evaluation by analyzing data in various scenarios. By simulating different scenarios, AI systems can help engineers understand how projects will perform under different conditions and make informed decisions about the best way to proceed. For example, in a manufacturing plant, engineers can use AI to simulate the impact of a supply chain disruption or a change in market demand, and adjust their planning accordingly. Scenario assessment enables companies to adapt quickly to changing circumstances and minimize the impact of unexpected events.

Strategic decision making and predictive analytics

At a more strategic level, AI has fundamentally changed the way business leaders approach planning and policymaking. Through predictive analytics, AI systems can identify market trends, changes in consumer behavior and growth opportunities. This provides business leaders with a clearer view of external and internal factors affecting the organization, enabling them to formulate more robust and forward-looking strategies (Brynjolfsson & McAfee, 2014).

Predictive analytics also enables companies to identify areas for improvement and plan for the long term. For example, in the energy sector, companies can use AI systems to forecast demand levels and adjust their production to optimize the consumption of natural resources and reduce costs. Similarly, in the healthcare sector, hospitals can use AI to forecast demand for medical services and plan their resources to avoid saturation and improve the quality of patient service (Russell & Norvig, 2021). AI, by providing detailed information about the likely future, enables leaders to make informed strategic decisions that position the organization to successfully meet upcoming challenges.

In addition, AI's ability to analyze large volumes of data and extract patterns and correlations not evident to humans enables the discovery of new business opportunities. By analyzing information about consumer behavior, purchasing preferences and usage patterns, companies can identify market niches and develop innovative products or services that respond to changing customer needs. This ability to anticipate market trends and adapt business strategies based on data provides a significant competitive advantage and enables organizations to stay relevant in a dynamic and constantly evolving business environment.

AI and ethics in decision making

While AI improves business decision making, it also introduces ethical challenges that require organizations to adopt transparency and accountability policies to address algorithmic bias and fairness in their applications. AI-based decision making can be influenced by algorithmic biases that affect the accuracy and fairness of recommendations. Biases in training data or algorithm settings can lead to decisions that favor or disfavor certain groups, generating ethical implications and potential legal issues. To address these challenges, it is critical that organizations implement policies for transparency and accountability in the development and use of AI (Daugherty & Wilson, 2018).

On the other hand, the use of AI in decision making also raises questions about responsibility for errors and unintended consequences. When a decision made by an AI system results in an error or an unfavorable outcome, it can be difficult to clearly assign responsibility. This lack of clarity in accountability poses both legal and ethical challenges, especially in sectors where decisions can have a significant impact on people's lives, such as healthcare or the financial sector. It is therefore critical that organizations develop governance and ethics frameworks for AI, ensuring transparency in their decision-making processes and auditability of AI systems (Russell & Norvig, 2021).

In conclusion, artificial intelligence transforms decision making at labor engineering, providing tools for data analysis, risk management and resource optimization. This advance redefines the role of engineers and establishes a dynamic where AI and human judgment collaborate for the benefit of organizational and ethical performance. Thanks to AI, organizations can proactively respond to operational challenges, improve their efficiency and strategically plan for the future. However, the use of AI in decision making also raises ethical and liability challenges that require a careful approach. As AI continues to evolve and expand its impact, companies must balance its benefits with the need to ensure responsible and transparent use of the technology. In this way, AI not only improves business decision making, but also redefines the role of engineers and leaders in creating an efficient, ethical and adaptive work environment.

3.3 AI as assistant and substitute in complex tasks

High-precision tasks, which previously required experts with specific training and experience, can now be performed or at least assisted by AI systems. This includes manufacturing processes, technical system diagnostics and quality monitoring, where AI can achieve levels of accuracy superior to the human eye. For example, in the semiconductor industry, where assembly and verification tasks require handling tiny components and identifying microscopic defects, AI has enabled automated inspection through computer vision and deep learning systems. These systems can analyze each component with millimeter accuracy, detecting defects that might otherwise go unnoticed by the human eye (Russell & Norvig, 2021).

Similarly, in medicine, AI has improved disease diagnosis through image analysis. Deep learning algorithms, trained on thousands of medical images, can identify patterns in X-rays, MRIs and CT scans with accuracy equal to or, in some cases, greater than that of specialized doctors. By detecting signs of diseases such as cancer in early stages, AI not only complements the work of healthcare professionals, but it acts as an invaluable tool in the diagnostic process, reducing margins of error and improving outcomes for patients (Topol, 2019).

Automation of complex data analysis and processing

Another of the most relevant applications of AI in work engineering is the automation of complex data analysis and processing. In sectors such as energy and logistics, the analysis of large volumes of data in real time is essential for fast and accurate decision making. AI makes it possible to analyze this data continuously, extracting valuable information that helps to optimize processes and improve efficiency. This type of analysis can include data on energy consumption, system behavior, and machinery operating conditions, allowing engineers to anticipate potential problems and make necessary adjustments to maximize performance (Brynjolfsson & McAfee, 2014).

In the energy sector, for example, AI helps monitor the performance of generation and distribution systems, identifying consumption patterns and optimizing resource use. AI systems can forecast energy demand based on variables such as weather conditions, consumption history and seasonal trends. This allows energy production to be adjusted according to predicted demand, avoiding waste and improving sustainability. In the logistics industry, AI is used to manage and optimize product movement and resource allocation, using real-time data to calculate optimal routes, reduce transportation costs, and improve on-time delivery (Acemoglu & Restrepo, 2018).

Simulation and optimization of complex systems

Simulation and optimization of complex systems is another task where AI has proven to be of great use in engineering work. AI makes it possible to create virtual models of complex systems and run simulations that replicate real conditions, providing engineers with the opportunity to test different scenarios and make informed decisions.

This type of simulation is particularly useful in industries where the margin for error is minimal and any failure can be costly or dangerous.

In the automotive industry, for example, manufacturers use AI to simulate the performance of their vehicles in different conditions, from crash safety simulations to performance tests in various weather conditions. These models allow adjustments to be made to the design and components before the vehicle reaches the production stage, reducing the risk of failure and optimizing the performance of the final product. This ability to foresee the impact of decisions before physically implementing them represents a significant competitive advantage by reducing costs and improving product quality without the need for extensive physical testing (Russell & Norvig, 2021).

Similarly, in the aerospace industry, AI simulation allows engineers to analyze aircraft structure and make design adjustments to optimize performance and safety. This type of simulation helps to identify problems early and improve component design, ensuring that the final product meets quality and safety standards. AI, by enabling accurate simulation of complex systems, not only improves the efficiency of design and production processes, but also reduces the risks and costs associated with prototyping and physical testing (Daugherty & Wilson, 2018).

Reducing errors and improving efficiency

One of the most prominent benefits of AI in the performance of complex tasks is the reduction of errors, which in turn increases efficiency and reduces costs. In processes that require a high level of accuracy, such as electronics manufacturing or drug delivery, AI systems can minimize the risk of human error. AI algorithms can monitor operations and detect anomalies that indicate potential errors, thus enabling operators to correct problems in a timely manner. This level of accuracy is especially important in sectors where errors can have serious consequences , such as in the pharmaceutical industry or medical device manufacturing (Topol, 2019).

For example, in the production of electronic devices, AI systems can analyze real-time data to ensure that each component meets quality standards. If a deviation in parameters is detected, the system can alert operators to make adjustments, thus preventing the production of defective products and reducing waste costs. In the healthcare sector, AI systems can review drug doses administered to patients and verify that prescriptions are followed correctly, helping to reduce medication errors and improving patient safety (Brynjolfsson & McAfee, 2014).

AI as a surrogate for specialized tasks

In some cases, AI not only assists engineers and technicians, but acts as a stand-in for specialized tasks. This is especially valuable in areas where the availability of experts is limited or where the task at hand is dangerous to humans. For example, in the mining industry, AI systems are used to operate equipment in hazardous environments, such as subway mines or oil rigs. These systems allow exploration and mining operations to be carried out without putting workers' lives at risk, reducing the impact of extreme environmental conditions on productivity (Frey & Osborne, 2017).

In the field of medicine, AI systems have also begun to play an important role in the area of assisted surgery. Surgical robots, controlled by AI, can perform operations with higher precision than human surgeons in certain interventions. This does not mean that AI replaces doctors, but rather that it acts as a complement that allows surgeons to perform complex procedures with greater precision and control. The collaboration between humans and machines in this context has led to a new form of medical care that combines the strengths of both, making it possible to provide high-quality treatment to patients (Topol, 2019).

AI and the development of technical competencies

The incorporation of AI into complex tasks has also changed the skills required in engineering work. As AI systems take on increasingly sophisticated responsibilities, engineers must acquire new skills to monitor, maintain and optimize these systems. This includes learning data analysis techniques, understanding machine learning algorithms, and the ability to interpret the results generated by AI systems. This change in the skills profile is critical to ensure that the implementation of AI maximizes its potential and minimizes the risks associated with its misuse (Russell & Norvig, 2021).

In addition, engineers must develop collaboration and adaptability skills, as AI works best in combination with human supervision. Rather than replacing workers, AI is presented as a tool that, under proper supervision, can perform complex tasks more efficiently and accurately. This symbiotic relationship between humans and machines raises the need for constant training so that engineering professionals are prepared to face technological challenges and take full advantage of AI capabilities.

AI has revolutionized the role of engineers in performing complex tasks, acting as an assistant and, in some cases, as a substitute in areas that require a high level of precision and specialization. The ability of AI to reduce errors, process large volumes of data and operate in hazardous environments has opened up new possibilities in engineering work, improving efficiency and safety in a variety of sectors. However, the successful implementation of AI requires engineers to develop new technical competencies and maintain active oversight over these systems, ensuring that AI fulfills its role as a valuable adjunct in the work environment.

3.4 Impact on productivity and organizational performance

One of the biggest impacts of AI on organizational productivity is in process automation. Thanks to AI, organizations can automate repetitive tasks that traditionally required human intervention, reducing production times and improving accuracy in the execution of activities. In the automotive industry, for example, AI-powered industrial robots assemble vehicles efficiently and with pinpoint accuracy, reducing the margin of error and ensuring that each component meets quality standards. This automation also allows human workers to focus on more strategic and creative tasks, thus increasing the added value of their work (Vera, 2021).

In a study by Perez and Martin (2019), it was found that the implementation of AI in manufacturing processes increased the productivity of companies by 25%, while reducing operating costs by 20%. These results evidence how process automation, facilitated by AI, allows optimizing the use of resources, decreasing waste and streamlining workflow, aspects that are fundamental to improve organizational performance. AI not only facilitates faster and more accurate execution of tasks, but also contributes to more efficient management of time and material resources, which represents a significant competitive advantage in highly demanding markets.

Cost reduction and resource optimization

Another important benefit of AI is the reduction of operating costs, which directly impacts the profitability of organizations. Through the use of machine learning algorithms and predictive analytics, AI systems can identify areas where costs can be reduced without affecting product or service quality. For example, in the energy industry, AI is used to optimize energy consumption in production plants by monitoring in real time and automatically adjusting consumption levels according to demand. This not only reduces energy expenditure, but also reduces the carbon footprint of organizations, contributing to greater sustainability (García & Fernández, 2020).

Resource optimization also applies to inventory management, where AI can forecast product demand and adjust stock levels accordingly. In the retail sector, companies such as Inditex have implemented AI systems to manage their inventories efficiently, reducing storage costs and minimizing the risk of losses from unsold products. This allows them to respond quickly and effectively to changes in demand, ensuring product availability in stores and improving customer satisfaction. According to a report by the consulting firm Deloitte (2021), companies that have adopted AI in inventory management have managed to reduce logistics costs by an average of 30%, improving profitability and operational efficiency.

Increased quality and process control

AI has also had a significant impact on improving the quality of products and services. Thanks to computer vision systems and real-time data analysis, organizations can monitor and control the quality of production processes with superior accuracy. In the pharmaceutical industry, for example, AI is used to analyze samples and detect possible contaminants in drugs, ensuring that each batch meets quality and safety standards. This type of automated control reduces the risk of human error and guarantees consistency in the quality of the final product (Sánchez & López, 2019).

A case study presented by Rodriguez and Molina (2020) in the food industry shows that the implementation of AI in quality control allowed a food processing company to reduce the number of defective products by 40%. This increase in quality control accuracy not only improves customer satisfaction, but also reduces costs associated with product returns and rework. In the field of work engineering, automated quality control facilitates constant monitoring of production standards, allowing engineers to identify and correct problems in a timely manner.

Improved customer experience

AI not only impacts an organization's internal processes, but also improves the customer experience by enabling faster and more personalized service. In the service sector, such as banking and e-commerce, AI is used to analyze customer preferences and behaviors, allowing companies to offer personalized recommendations and services tailored to their specific needs. According to a study by Gonzalez and Ortiz (2019), companies that use AI to personalize the customer experience have been able to increase customer retention by 15% and increase revenue by 10%. This type of personalization not only improves customer satisfaction, but also strengthens brand loyalty and contributes to the sustained growth of the organization.

For example, in e-commerce, AI algorithms analyze customers' browsing and purchase history to provide relevant product recommendations, increasing the likelihood of conversion and creating a more engaging and efficient shopping experience. These types of strategies allow companies to differentiate themselves from their competitors and build stronger relationships with their customers. In addition, AI enables automated customer service through chatbots and virtual assistants, which provide immediate answers to common queries, improving service efficiency and user satisfaction (Vera, 2021).

Impact on the organizational culture and adaptation to change.

The implementation of AI in organizations has also had an impact on organizational culture by driving greater adaptation to change and fostering a culture of innovation. As AI takes over repetitive and operational tasks, employees can focus on more value-added activities, such as strategic planning, problem solving and developing new ideas. This has led to a change in the perception of work, as employees now see AI as a tool that facilitates their work rather than a threat to their jobs (Garcia & Fernandez, 2020).

In addition, the adoption of AI in organizations requires a culture of continuous learning, where employees are willing to acquire new skills and adapt to technological changes. This transformation in organizational culture is critical for companies to take full advantage of AI capabilities and ensure a successful implementation. According to Perez and Martin (2019), companies that have adopted a culture of innovation and continuous learning have experienced an increase in employee satisfaction and engagement, which contributes to better organizational performance and a more collaborative and adaptive work environment.

Challenges of AI in organizational performance

While the benefits of AI on productivity and organizational performance are evident, its implementation also presents challenges. One of the main obstacles is the initial cost of investing in AI technologies and the infrastructure required to operate them. Implementing AI systems involves significant investment in hardware, software, and staff training, which can be challenging for smaller or resource-constrained companies (Cano et al., 2020).

Another challenge is resistance to change from employees, who may see AI as a threat to their jobs. To overcome this resistance, it is critical that organizations implement communication and training policies that explain the benefits of AI and train employees in the use of these technologies. Creating a work environment in which employees perceive AI as a supportive tool, rather than a replacement, is key to ensuring successful implementation and maximizing organizational performance (Garcia & Fernandez, 2020).

Artificial intelligence has had a profound impact on productivity and organizational performance, enabling process automation, resource optimization and improved product and service quality. These improvements not only contribute to greater profitability, but also strengthen the competitive position of organizations in the marketplace. However, the implementation of AI also poses challenges, such as the need for a culture of continuous learning and investment in technology infrastructure. As organizations move forward with AI adoption, it is critical that they focus on maximizing the benefits of this technology while addressing the challenges associated with its implementation. AI, when applied strategically and ethically, can become a competitive advantage that transforms organizational performance and work culture in a sustainable way.

Chapter 4. Skills of the Future: Human Adaptation to the Technological Revolution

Technological transformation in the age of artificial intelligence (AI) and automation has reshaped the requirements and expectations of the work environment. Skills that were once indispensable are being replaced or supplemented by new competencies that enable professionals to collaborate effectively with technology. This chapter explores in depth the skills that will enable the workers of the future to live and thrive in a world where AI and technology play increasingly important roles.

First, soft skills - such as creativity, critical thinking and collaboration - have gained significant value. These human skills are not only still relevant, but are now considered critical in an environment where machines are taking over routine or analytical tasks. According to the World Economic Forum (2018), 85% of the jobs that will exist in 2030 have not yet been created; moreover, many of these roles will require interpersonal and communication skills, areas in which machines cannot yet compete with humans.

Creativity emerges as one of the most valuable skills in an AI-driven environment. While machines are proficient at processing data and performing complex calculations, it is the creative ability of humans that enables the development of innovative and disruptive solutions. McKinsey & Company (2018) highlights that creative skills will be essential to tackle complex problems that require novel approaches, as human creativity remains a key advantage in a highly automated environment.

Critical thinking is another crucial competency in the information age. With access to massive volumes of data, professionals must evaluate and analyze such information to make informed decisions. Brown et al. (2020) emphasize in their Harvard Business Review study that critical thinking will be one of the most sought-after skills leaders of the future, as the ability to evaluate and discern continues to be a distinctive human quality in the face of technology.

In addition, today's work environment demands effective collaboration between humans and machines. This integration does not imply substitution, but rather a synergy in which employees learn to work together with AI systems to optimize productivity. Deloitte (2021) identifies human-AI collaboration as a key competitive differentiator for organizations of the future, as it enables data analytics to be combined with human intuition to solve problems efficiently and accurately.

Adaptability has also become an essential skill in the digital age. As technology evolves, adaptability allows professionals to stay current and meet the challenges arising from rapid changes in the marketplace. Studies conducted by PwC (2021) highlight that employees who demonstrate flexibility and willingness to learn new technologies are the most valued in today's corporate environments.

Adopting a continuous learning mindset is equally essential to cope with technological change. The speed of innovation demands that workers acquire new knowledge on an ongoing basis to stay relevant. Bersin by Deloitte (2019) concludes that lifelong learning is a hallmark of successful employees in high-tech environments, as those who constantly train are more likely to adapt and thrive.

Demand for technical skills, such as programming, data analytics and data science, is also booming. These competencies are essential for those looking to work at the intersection between humans and technology. Programming knowledge, for example, enables professionals to develop and customize technology solutions, while data analytics skills facilitate the interpretation of large volumes of information to make strategic decisions. The World Economic Forum (2020) forecasts that the demand for these technical skills will increase considerably in the coming years.

Within this context, continuous education and training are essential. Companies have a responsibility to provide their employees with the necessary tools to meet the challenges of the AI era. By implementing training and development programs, organizations not only improve the technical competence of their workers, but also promote a culture of resilience and adaptation. A report by Accenture (2020) highlights that companies that invest in the development of their staff are the ones that are best able to adapt to market changes.

Changes in job skills also have a profound impact on organizational structure. In an environment where machines automate routine tasks, employees can focus on strategic activities that require a high level of analysis and creativity. PwC (2021) argues that organizations that manage to effectively integrate technology and human talent perform better in innovation and customer satisfaction.

In this context, a new discipline emerges: work engineering. This discipline advocates redesigning organizational tasks and processes to maximize the use of technology and human competencies. According to Accenture (2020), work engineering enables organizations to restructure themselves to adapt to digital transformation, improving efficiency and enabling greater collaboration between humans and machines.

Emotional management skills also become relevant. Living with advanced technology can be stressful for some employees, who fear the loss of their jobs in the face of advancing automation. The World Economic Forum (2021) suggests that emotional management skills will be key to maintaining well-being in the workplace and reducing anxiety caused by technological change.

Organizational leaders have a crucial role to play in this transition process. They need to develop competencies in adaptive leadership and hybrid team management, integrating human workers and technology. Harvard Business Review (2020) suggests that leaders with digital team management skills can facilitate better adaptation to change and ensure that their teams feel empowered and valued in an increasingly digital environment.

Ethical values and responsibility in the use of AI are also emerging areas of competence. As AI takes on decision-making tasks, it is essential that workers understand the ethical principles governing the use of this technology. According to the AI Ethics Institute (2020), training in technology ethics will enable professionals to make decisions that promote the common good and avoid bias or prejudice.

Finally, it is important to note that this technological revolution poses both challenges and opportunities. Task automation can free employees from repetitive work, allowing them to engage in activities that require creativity, innovation and empathy. Deloitte (2021) concludes that workers in the most innovative organizations see AI as an opportunity to expand their competencies and take on more fulfilling roles.

4.1. Soft skills in the age of AI

The integration of artificial intelligence (AI) into the workplace has underscored the relevance of soft skills, which include competencies such as creativity, critical thinking, empathy and communication. In an age where machines can perform repetitive tasks and process large amounts of data, human skills are indispensable to provide differential value. According to the World Economic Forum (2018), these soft skills are among the most valued skills in the workforce of the future, as they complement the technological capabilities of AI and enable workers to perform roles that require greater adaptation.

Creativity has become a decisive factor in organizations. While AI and automation enable process efficiencies, it is human creativity that drives innovation and enables companies to differentiate themselves in an increasingly competitive market. According to a study by McKinsey & Company (2018), creativity will be critical to solving complex problems in an automated environment, as original ideas can only emerge from human ingenuity, not preset algorithms.

Critical thinking is another competency that has gained relevance in this era of massive data. The ability to analyze and question information enables employees to make informed decisions, a crucial aspect when companies face data overload and need to assess its veracity and applicability. Brown et al. (2020), in an analysis for Harvard Business Review, note that critical thinking is essential to effectively filter and synthesize information, particularly when AI is used to generate real-time data.

Empathy and communication skills are also essential in a digitized work environment. AI can facilitate communication, but it cannot replicate the emotional connection and deep understanding that characterizes human relationships. Deloitte (2021) argues that empathy and active listening skills are critical skills for leading diverse teams and maintaining organizational cohesion in an increasingly virtual world.

Collaboration in a human-IA team requires change management skills and adaptability. Instead of seeing AI as a threat, employees must learn to work alongside it to maximize their efficiency and creativity. According to the World Economic Forum (2020), companies that foster a culture of human-machine collaboration achieve significantly improved results, as employees focus on value-added tasks while machines execute repetitive processes.

Adaptive leadership stands out as an essential soft skill for professionals looking to lead teams in a highly technological environment. In the age of AI, leaders must be flexible and foster an environment of trust and transparency. Harvard Business Review (2021) highlights that adaptive leaders are critical to facilitating the integration of technology into organizations, as their emotional management skills and adaptability help mitigate resistance to change.

Teamwork and collaboration are essential for teams to integrate AI technologies effectively. PwC (2021) argues that teams that master collaboration and communication skills are the most successful in AI implementation, as they achieve better harmony between technology and human processes. Effective interaction between teams is key to ensuring that AI serves its purpose without undermining the value of human capital.

Mental flexibility and the willingness to learn have also come to the fore. Rapidly evolving technology requires employees to be able to constantly adapt to new tools and processes. Bersin by Deloitte (2019) emphasizes that a continuous learning mindset is crucial to cope with the demands of digital transformation and that those workers who regularly update themselves are more likely to excel.

In terms of innovation, creative thinking becomes an asset that drives companies to position themselves as leaders in their industries. Accenture (2020) suggests that human creativity is key to harnessing AI capabilities in innovative and relevant ways, enabling companies to develop solutions that address complex problems with a unique and competitive approach.

The ability to communicate clearly and effectively is also imperative, especially in roles that require the interpretation and transmission of AI-generated data. According to Harvard Business Review (2020), the ability to explain technical information in an understandable way is critical for other departments to understand and use the information properly. Lack of clear communication can lead to poor decisions and lack of cohesion within the organization.

Conflict resolution is a soft skill that takes on a relevant role in the age of technology. As AI replaces some functions, tensions and fears arise in the workforce, requiring negotiation and conflict resolution skills. The AI Ethics Institute (2021) highlights that the ability to mediate complex situations is crucial to maintaining organizational well-being and employee morale in times of change.

Developing competencies in emotional intelligence and empathy benefits teams, especially when faced with technological change. Emotional intelligence enables leaders and team members to better understand and manage their own emotions and those of their colleagues, an essential aspect in times of uncertainty. According to the World Economic Forum (2021), emotional intelligence skills are critical to reducing work stress and improving employee engagement in changing environments.

4.2. Continuous adaptation and lifelong learning

Continuous adaptation and lifelong learning have become essential elements in the current context of digital transformation. The rapid advancement of technology and the constant introduction of innovations require workers to acquire a flexible and constant learning mindset. According to the World Economic Forum (2020), 50% of all employees will need to acquire new skills by 2025, underscoring the need for continuous updating to remain competitive.

Lifelong learning involves more than occasional training; it is a proactive approach that enables employees to anticipate and adapt to the demands of the labor market. PwC (2021) explains that a continuous learning mindset is crucial for those who wish to remain relevant in a rapidly changing environment that demands skills that evolve along with technology.

The adoption of AI and automation in the workplace has transformed competency demands, requiring employees to not only master technical skills, but also develop the ability to learn new tools and processes. According to Bersin by Deloitte (2019), the ability to learn quickly and adapt to new technologies is a characteristic of employees who manage to excel in digitized environments, which becomes a significant competitive advantage.

Lifelong learning enables professionals to adapt to a world of work where skills obsolescence is increasingly prevalent. Harvard Business Review (Brown et al., 2020) argues that those who adopt a constant learning approach are more likely to remain competitive in their industries and successfully meet market challenges in the digital age.

Access to information and online education has made it easier for employees to train themselves. Digital education allows workers to access courses and resources from anywhere, reducing barriers and facilitating the acquisition of new skills. Accenture (2020) points out that online learning has become a key tool for skills development, as allows employees to improve their skills at their own pace and according to their specific needs.

Adaptability, in this context, has become an invaluable skill. The ability of workers to modify their approaches and strategies according to new market demands and technological advances is fundamental for the sustainability of companies. Deloitte (2021) stresses that organizations that foster a culture of adaptation are able to respond more agilely and effectively to changes in the environment, which positions them favorably against the competition.

As automated tasks reduce the need for certain traditional skills, employees must reinvent themselves. Mental flexibility and a willingness to acquire new knowledge have become crucial to ensure a stable career path. The AI Ethics Institute (2021) asserts that lifelong learning is critical for employees to find opportunities in a marketplace where technical skills are rapidly being replaced by more advanced ones.

In the context of automation, continuous learning not only benefits employees, but is also a strategic resource for organizations. PwC (2021) concludes that companies that invest in employee training have a competitive advantage in the digital era, as they achieve greater resilience in the face of change and reduce the risk of skills obsolescence among their staff.

Investing in skills development is also a strategy for retaining talent. Those companies that offer opportunities for growth and learning tend to generate greater engagement and loyalty among their employees. McKinsey & Company (2018) suggests that a culture of continuous learning not only enables companies to meet the challenges of digitalization, but also contributes to the motivation and satisfaction of their employees.

4.3. In-demand technical skills: programming, analytics and data science

The increasing adoption of artificial intelligence and automation in organizations has boosted the need for specific technical skills, such as programming, data analytics and data science. These competencies enable professionals to work at the intersection between humans and technology, enhancing their abilities to interpret, build and collaborate with automated systems. According to the World Economic Forum (2020), 84% of companies are implementing AI or exploring advanced technologies, highlighting the demand for technical skills across multiple sectors.

Programming is one of the most sought-after technical skills in today's market, as it allows employees to develop customized technology solutions and tailor AI systems to the specific needs of their organizations. A report by PwC (2021) reveals that knowledge in programming languages such as Python, R and JavaScript is critical for professionals who want to excel in high-tech sectors, especially in AI and automation development positions.

Data analytics capabilities are equally essential in the digital age, where data is considered one of the most valuable assets of organizations. Data analysts are responsible for extracting meaningful insights from large volumes of data, enabling businesses to make informed decisions. According to Harvard Business Review (Brown et al., 2020), data analysis has become a critical skill for strategic decision making, as it allows identifying patterns and trends in consumer behavior and optimizing internal operations.

Data science, which includes advanced modeling and machine learning techniques, is a rapidly growing discipline that enables businesses to take full advantage of AI. According to Accenture (2020), data scientists are essential for developing predictive models and advanced algorithms that can be applied to a variety of business functions, from marketing and supply chain management to customer service and financial forecasting.

The combination of programming and data science skills enables professionals to work on creating AI systems that optimize business processes. The Massachusetts Institute of Technology (MIT) (2021) suggests that employees with programming and data skills are able to develop customized algorithms and optimize operations, which is critical in a highly competitive environment.

Knowledge of data visualization techniques is also essential for data to be presented in an understandable and attractive way. Data visualization experts transform large volumes of data into graphs and charts that facilitate the interpretation of information. According to Deloitte (2021), the ability to visualize data effectively helps employees and leaders better understand performance metrics and make evidence-based strategic decisions.

Artificial intelligence applied to data management has generated an increase in demand for experts in machine learning and data mining. These skills enable workers to extract useful information from unstructured and structured data, which optimizes processes and improves decision-making capabilities. McKinsey & Company (2021) highlights that machine learning is becoming one of the most valued areas within data science due to its ability to automate tasks and predict future behaviors.

Professionals with data science and analytics skills benefit from a holistic perspective on projects, as they have the ability to understand and manipulate both the technical infrastructure and the strategic objectives of the organization. According to the World Economic Forum (2020), workers with specialized technical skills are more valuable to companies, as their ability to connect technology with business strategies allows them to add significant value.

The handling of databases and data management systems is another in-demand technical skill, as it allows professionals to organize and maintain large amounts of information efficiently. Knowledge in SQL and other database management systems has become indispensable for those working in sectors such as finance, healthcare and e-commerce, where data management is critical for decision making (PwC, 2021).

Automating processes through AI requires employees to have skills in programming and understanding algorithms to design and adapt systems that increase efficiency. A Harvard Business Review report (2020) highlights that skills in algorithms and automation enable employees to reduce the time it takes to execute repetitive tasks, thereby improving productivity at all levels of the organization.

The use of advanced data analytics platforms, such as Hadoop and Spark, has become a valuable competency for those who wish to work in big data management. The ability to process large-scale data is crucial for companies seeking to better understand their customers and optimize their operations, especially in industries that rely on real-time data analytics (Deloitte, 2021).

4.4. Education and training for the new labor engineering

The digital revolution has transformed not only the skills required, but also the ways in which organizations train and prepare their employees. Continuous training is essential for workers to adapt to change and for companies to remain competitive. According to Accenture (2020), training in technical competencies and soft skills is key for companies to take advantage of artificial intelligence without jeopardizing their organizational culture.

Training in technical and programming skills has gained importance, especially in companies seeking to implement AI in their processes. PwC (2021) suggests that technology skills training programs should be aligned with the company's business strategies, enabling employees to acquire competencies directly applicable to their roles.

The development of data science and analytics skills is also a priority in training initiatives, as it enables employees to understand and apply data in decision making. According to Deloitte (2021), companies that implement data science training programs achieve better results in terms of efficiency and quality, as their employees are able to transform complex data into useful information.

Soft skills training is equally relevant in the digital age. Communication, collaboration and conflict resolution skills are essential for employees who must work alongside technology and adapt to changing environments. The World Economic Forum (2020) highlights that soft skills are as important as technical skills, as the ability to interact and collaborate effectively in human-IA teams largely determines the success of advanced technology projects.

The new engineering of work involves a restructuring of roles and processes to maximize the potential of AI and technology. Companies must prepare their employees not only in technical skills, but also in change management and adaptive leadership capabilities. Harvard Business Review (2021) suggests that adaptive leadership is essential in environments that are constantly evolving, as it enables leaders to guide their teams in integrating new technologies and processes.

Education in AI ethics and digital responsibility is another crucial component of today's training. Employees need to understand the ethical and social implications of the technology they develop and use. According to the AI Ethics Institute (2021), training in ethical principles enables workers to address challenges such as privacy and bias in algorithms, contributing to a more responsible use of technology.

Creating a culture of continuous learning is key to address the rapid obsolescence of skills. Accenture (2020) highlights that those companies that foster a constant learning mindset in their employees are more successful in adapting to digital transformation, as they manage to keep up with new trends and market demands.

The use of online learning platforms allows employees to access training flexibly and on an as-needed basis. According to McKinsey & Company (2018), online learning has become an essential resource for workers to acquire skills in a constantly changing environment, allowing them to update their skills and respond to changes in a timely manner.

Corporate training initiatives also include training in data management, a key skill in the information age. The ability to manage data efficiently enables companies to optimize their operations and gain valuable insights. Deloitte (2021) suggests that employees trained in data management and analytics can make informed decisions, thereby increasing enterprise value.

Investing in training also promotes talent retention and employee loyalty. Those organizations that provide opportunities for development and growth tend to see an increase in employee engagement. According to PwC (2021), employees who receive ongoing training feel more valued and committed to their employers, which benefits the organization in terms of productivity and stability.

Customization of training programs is another important trend. Harvard Business

Review (2020) argues that training programs tailored to the specific needs of each employee are more effective, as they allow workers to develop skills relevant to their roles and responsibilities. This is especially useful in large companies seeking to maximize the impact of their training investments.

4.5. Ethics and Responsibility in the Age of Automation

The integration of automated technologies into the work environment raises a number of ethical dilemmas that cannot be ignored in the age of artificial intelligence (AI). With the increasing implementation of automation in decision-making processes, data analysis and other critical tasks, it is imperative to consider the associated ethical issues. Ethics in AI involves both responsible design of technologies and conscious decision making that prioritizes human welfare. According to the AI Ethics Institute (2021), workers and leaders must be trained to understand and manage the ethical implications of AI, preventing the technology from causing negative effects on society.

Corporate social responsibility (CSR) is an essential element in the implementation of automated technologies. Companies must take responsibility for the social and labor impacts arising from automation by committing to using technology in a fair and equitable manner. A report by Accenture (2020) points out that CSR in the age of AI includes not only minimizing risks to employees, but also developing policies that ensure that the benefits of technology are distributed equitably, protecting the most vulnerable workers.

Transparency in the use of AI is another fundamental aspect of ethics in automation. Decisions generated by automated systems must be understandable and traceable, especially in sectors where they directly affect employees and consumers.

Harvard Business Review (2021) suggests that transparency in AI algorithms and processes is key to gaining the trust of users, who must understand how and why certain decisions are made. Lack of transparency in automation can lead to bias and a lack of accountability in organizations.

Algorithmic bias is one of the most important ethical concerns in the age of automation. As AI systems make decisions based on large volumes of data, there is a risk that they will perpetuate bias or discriminate against certain groups. McKinsey & Company (2021) warns that algorithms trained on historical data tend to reflect existing biases in society, which can result in unfair outcomes. Organizations should be aware of this risk and should employ practices to mitigate bias in their automated systems.

Data privacy is also a critical concern in automation. The collection and analysis of large volumes of personal data requires companies to take steps to protect information and ensure that data is used ethically. According to Deloitte (2021), data privacy must be a priority for organizations implementing AI, as a failure to protect data can affect a company's trust and reputation. Clear privacy policies and regulatory compliance are critical in this regard.

Ethical decision making in an automation environment requires a combination of technical skills and a strong commitment to ethical principles. Employees must be aware of the implications of their decisions and how these may affect society as a whole. The AI Ethics Institute (2021) suggests that companies adopt an ethical approach to the design and use of AI, incorporating principles such as fairness, transparency, and accountability into their automated processes.

The impact of automation on employment also raises ethical questions. The implementation of technology may lead to the elimination of certain roles, affecting the workers who depend on these jobs. PwC (2021) stresses that companies have a responsibility to prepare their employees for the transition to a more automated work environment by providing training and reassignment opportunities. This social responsibility helps mitigate the negative impact on workers and promotes a fairer transition to automation.

Ethics in AI also extends to the design and development of AI systems, including ensuring that systems are safe and do not cause unintended harm. Harvard Business Review (2021) argues that AI development must follow ethical standards that minimize risks and protect the integrity of users. Ethics in design is a responsibility that engineers and developers must assume to avoid negative consequences in society.

Responsible use of AI in automated decision making requires adequate human oversight to avoid errors or abuses. Automated decisions cannot be absolute; human professionals need to oversee these systems to ensure that decisions are fair and respect human rights. Deloitte (2021) argues that human oversight is key to ensuring that automated decisions align with the organization's ethical values and do not cause harm to employees or customers.

Organizations should implement audit mechanisms to assess the performance and fairness of AI systems. These audits can identify potential bias, privacy, or transparency issues, which helps companies adjust their systems to meet ethical standards. A McKinsey & Company (2021) study highlights that regular audits are essential to ensure that AI systems comply with ethical regulations and do not generate harmful outcomes.

AI ethics also includes the principle of harm minimization, which urges companies to avoid using AI in situations where it may cause a negative impact on individuals or society. According to the AI Ethics Institute (2021), harm minimization is a fundamental principle for the ethical use of AI, as it ensures that the technology is used with a focus on social welfare and not just economic benefit.

Finally, ethics and responsibility in the age of automation require an organizational commitment to training in ethical principles and responsibility at all levels. This involves not only training employees and leaders in digital ethics, but also establishing organizational policies that foster an ethical culture in the use of technology. Accenture (2020) concludes that companies that take a proactive ethical stance in their use of AI achieve greater acceptance and trust from their employees and customers, which is critical in an increasingly ethically conscious marketplace.

Conclusion

The technology revolution has redefined the competencies needed to thrive in today's workplace, highlighting the importance of both soft and technical skills in a context of rapid automation and artificial intelligence. Soft skills-such as creativity, critical thinking and empathy-enable employees to bring human value that technology cannot replicate. These skills are essential for innovation and complex problem solving, aspects that require human intuition, imagination and judgment. In an environment where machines perform data analysis at high speed, soft skills complement technology by providing the perspective and discernment needed to interpret and apply results in a meaningful way (World Economic Forum, 2018).

Continuous adaptation and lifelong learning are now indispensable as technology evolves at a dizzying pace. Professionals need to adopt a constant learning mentality, updating their skills so as not to fall behind. This not only benefits employees, who can stay competitive and relevant, but also companies, which gain in resilience and agility. Organizations that foster a culture of continuous learning and development manage to retain talent and are better able to adapt to market changes (Bersin by Deloitte, 2019; PwC, 2021).

Likewise, technical skills in programming, analytics and data science have become essential for work at the intersection of humans and technology. These competencies enable professionals to not only handle large volumes of data and create AI-based solutions, but also understand and control the tools that facilitate growth and efficiency in their companies. In sectors such as finance, healthcare and technology, the demand for these skills continues to increase as organizations seek to leverage data and automated systems to make strategic decisions (World Economic Forum, 2020; Accenture, 2020).

Education and training are crucial to effectively integrate workers into this new paradigm. Companies that invest in training programs that encompass both soft and technical skills have a competitive advantage, as they develop a workforce that is prepared to interact with technology in a productive and ethical manner. Training in areas such as AI ethics, adaptive leadership and data management enables companies to promote a responsible and innovative culture. By doing so, organizations not only optimize their operations, but also ensure greater cohesion and engagement in their workforce (Deloitte, 2021; Harvard Business Review, 2020).

In conclusion, the work environment of the future demands a balanced combination of soft skills, technical skills and a lifelong learning mentality. As technology continues to transform work processes and structure, professionals must adapt and evolve in parallel. This chapter demonstrates that the true value of technology is realized when it is complemented by human capabilities, creating a synergy that empowers both organizations and individuals.

Chapter 5: The future of work: challenges and opportunities in the age of digitalization.

The era of digitalization has profoundly redefined the employment landscape, transforming the nature of work, business models and the dynamics between employers and employees. The incorporation of advanced technologies, such as artificial intelligence (AI), machine learning and process automation, has opened up a new world of possibilities, but also poses significant challenges for the future of work. While, on the one hand, digitalization has made it possible to optimize efficiency, reduce costs, and improve productivity in various industries, it has also generated uncertainty and challenges around employment and the nature of the skills required in the work environment. This chapter explores how digitalization impacts current business models, and examines the challenges and opportunities that arise as work continues to evolve.

Digitalization is driving radical changes in business models, which are now forced to adapt to a reality where innovation and agility are critical factors for success. As technology redefines business relationships, companies are finding it necessary to adjust their structures and strategies to respond to the demands of a market that is advancing at a dizzying pace. The emergence of digital business models, such as collaborative economy platforms, e-commerce and AI-based solutions, has changed not only how companies generate revenue, but also how they interact with their customers, manage their operations and face competition. An organization's ability to adapt and evolve in this digital environment has become a crucial factor for its long-term sustainability (García & Fernández, 2020).

In this context, both challenges and opportunities arise for companies and workers. Digitalization poses challenges in terms of employability, as many manual or routine tasks are being replaced by automated systems and robots. According to a study by the World Economic Forum (2020), it is estimated that by 2025 automation will have replaced approximately 85 million jobs worldwide, although new roles focused on technology and innovation are also expected to be created. These changes generate a labor landscape in which adaptability and continuous learning become essential competencies for the workforce.

Digitalization affects not only existing jobs, but also the skills that workers need to thrive in the workplace. With the advent of AI and other advanced technologies, technical skills such as programming, data analysis and technology systems management have become increasingly common requirements in many sectors. However, it also highlights the importance of "human" skills, such as creativity, problem solving and emotional intelligence, which complement the work performed by machines and enable employees to play a strategic role in their organizations. In this sense, digitalization does not only imply the automation of tasks, but also a revaluation of human skills in the workplace (Cano et al., 2020).

Another important challenge of digitization is the need to establish adequate regulations to ensure data privacy and information security. In an increasingly digitized work environment, the handling of personal and business data has become common practice, and the protection of this information is a key concern. Organizations must comply with privacy rules and regulations, such as the General Data Protection Regulation (GDPR) in Europe, to ensure that the use of technology respects the privacy rights of their employees and customers. This implies an investment in cybersecurity and the creation of data protection policies that adapt to the demands of the digital environment.

Despite these challenges, digitalization also offers multiple opportunities for companies and workers who know how to take advantage of it strategically. Technology enables the creation of new business models that not only improve efficiency, but also open doors to new markets and innovative ways of generating value. For example, models based on the collaborative economy, such as Uber and Airbnb, have emerged thanks to digital technology, allowing companies to directly connect users with services and goods through online platforms. These digital business models have not only changed the structure of sectors such as transportation and hospitality, but have also created flexible job opportunities for people seeking freelance or part-time work options (Vera, 2021).

In the work environment, digitalization has led to the flexibilization of schedules and the implementation of remote work. The COVID-19 pandemic accelerated this trend, and many companies were forced to adopt telecommuting as a measure to ensure the continuity of their operations. This experience demonstrated that many tasks can be performed remotely, offering benefits to both employees, in terms of work-life balance, and employers, who can reduce operating costs and expand their search for talent beyond geographic limitations. Telework and hybrid work models are trends that are likely to continue in the future, generating a significant transformation in the work environment and work dynamics (González & Ortiz, 2019).

On the other hand, digitalization also raises questions about work ethics and equity. Automation and AI present the risk of deepening labor inequalities by concentrating demand on highly skilled professionals and reducing opportunities for those workers with less technical training. To address these challenges, it is critical that organizations and governments work together to create digital inclusion policies and training programs that enable all workers to adapt to the new demands of the digital environment. In this way, it can be ensured that digitalization translates into equitable benefits and sustainable development (Pérez & Martín, 2019).

In short, the future of work in the era of digitalization presents both challenges and opportunities. Adaptability and readiness to change will be determining factors for companies and workers to reap the benefits of technology and respond to the demands of a constantly evolving labor market. In this chapter, we will explore the changes in business models driven by digitization, as well as the challenges and opportunities that these changes generate. From the reorganization of business structures to the creation of new ways of working and the need to develop digital competencies, the future of work will largely depend on how organizations and individuals adapt and adopt technological tools responsibly and strategically.

5.1 Transformations in business models

Digitalization is generating a profound transformation in business models, changing the way companies operate, interact with their customers and generate value. In an environment where technologies are advancing rapidly, organizations are faced with the need to adapt in order to remain competitive and take advantage of the opportunities offered by the digital era. This transformation encompasses both the internal restructuring of companies and the implementation of strategies that integrate disruptive technologies such as artificial intelligence (AI), big data, automation and the cloud. As the digital world continues to evolve, traditional business models are being challenged by new approaches that offer greater efficiency, flexibility and responsiveness to customer demands.

Digitalization and Innovation in Business Models

Digitalization has facilitated the creation of business models that not only innovate the way products and services are delivered, but also transform the relationship between companies and consumers. In the past, business models used to be based on static and linear processes, where the production and distribution chain followed a rigid structure. However, with the rise of digital technology, this structure has given way to dynamic, agile and customer-centric models (World Economic Forum, 2020).

One of the clearest examples of this change is the emergence of platform-based business models. Companies such as Amazon, Uber and Airbnb have changed the way consumers access products and services, using digital platforms that connect users with providers efficiently and without the need for traditional intermediaries. These platforms work through the intensive use of data and AI algorithms that optimize the user experience, personalize offers and adjust supply to demand in real time (González & Ortiz, 2019). These types of models have enabled companies to expand rapidly and operate in a global market, while reducing

operating costs and increase their reach.

Personalization is another key element in digital business models, as it allows companies to tailor their products and services to the individual needs of consumers. Digital platforms employ data analytics tools to gather information about user behavior and preferences, enabling companies to deliver more personalized experiences. This has led to a transformation in sectors such as retail, where companies such as Netflix and Spotify offer personalized recommendations based on each user's interests, creating a closer and more personalized relationship with their customers (Vera, 2021).

Automation and process optimization

Automation is one of the pillars of digital transformation in business models, as it allows companies to optimize their operations and reduce costs by implementing technologies that perform tasks autonomously. Robotics, artificial intelligence and process automation through algorithms have been essential in the reconfiguration of sectors such as manufacturing, logistics and financial services. In the manufacturing industry, for example, automation has made it possible to reduce production times, minimize the margin of error and improve product quality. The ability of robots to perform repetitive and high-precision tasks allows companies to improve their efficiency and respond more quickly to market demands (García & Fernández, 2020).

In financial services, automation has transformed the way companies manage their processes and services. Through the implementation of algorithms and AI systems, financial institutions can automate tasks such as identity verification, fraud detection and real-time transaction processing. This not only improves efficiency, but also reduces costs associated with manual processing and minimizes human error, resulting in greater reliability for users and an optimization of time and resources (Sanchez & Lopez, 2019).

Automation has also facilitated the analysis of large volumes of data, allowing companies to make adjustments to their operations based on the information obtained. Data analysis has become a fundamental tool in business decision making, enabling leaders to identify patterns and trends that facilitate strategic planning. With the ability to process data in real time, companies can respond in an agile manner to changes in the environment, adjusting their business models to adapt to market needs and anticipate potential challenges (Pérez & Martín, 2019).

Business models based on the collaborative economy

The collaborative economy is a business model that has emerged from digitalization and allows people to share or rent resources through digital platforms, rather than own them. This model has changed the dynamics of ownership and consumption in sectors such as transportation and hospitality, where companies such as Uber and Airbnb have created platforms that connect users with service providers quickly and conveniently. The collaborative economy is based on the idea of maximizing the use of existing resources, allowing people to access goods and services without the need to acquire them permanently (González & Ortiz, 2019).

The popularity of the collaborative economy reflects a trend toward sustainable consumption and reduced waste of resources. Collaborative platforms allow consumers to enjoy greater flexibility and convenience, while service providers can monetize their assets and earn additional revenue. This model not only transforms the relationship between consumers and providers, but also has a positive impact on the economy by encouraging the efficient use of resources and reducing the need for additional production (World Economic Forum, 2020).

In addition, the collaborative economy has promoted a shift in consumer preferences , where users value experience and access more than ownership of goods. This trend has led to a shift in the way companies design their business strategies, focusing on creating engaging user experiences and developing intuitive and efficient platforms. The collaborative economy has proven to be a flexible and adaptable business model, responding to the demands of an increasingly digitized and sustainability-conscious market (García & Fernández, 2020).

Adaptation and organizational agility in the digital era

Digital transformation has forced companies to adapt quickly and develop a culture of organizational agility that allows them to respond to changes in the environment efficiently. In the digital era, organizations must be able to identify opportunities and react to technological and market challenges quickly, which requires a flexible organizational structure and a change-oriented mindset. Companies that have successfully embraced digitization tend to be those that have invested in innovation and foster a culture of continuous learning, where employees are motivated to acquire new skills and adapt to emerging technologies (Vera, 2021).

Organizational agility also implies the ability to implement digital transformation strategies in a progressive manner, adapting to customer needs and the specific context of the company. This requires business leaders to understand the importance of technology in creating value and improving internal processes, and to be able to coordinate efforts to integrate digitization at all levels of the organization. The adoption of technologies such as big data and predictive analytics allows companies to make strategic adjustments based on concrete data, which facilitates informed decision-making and reduces the risk of errors (Pérez & Martín, 2019).

In this sense, companies must be able to adapt to digital business models without losing their corporate identity and without neglecting their core values. Digitization does not imply the elimination of traditional processes, but the integration of technological tools that complement the organization's existing strengths . Those companies that manage to balance tradition with innovation are best positioned to face the challenges of a changing and competitive environment.

Challenges in the implementation of digital business models

While digitization presents multiple opportunities for the transformation of business models, it also poses significant challenges that companies must overcome to ensure successful implementation. One of the biggest hurdles is the initial investment in technology infrastructure, which can be costly and complex to implement, especially for small and medium-sized enterprises. The adoption of advanced technologies, such as AI and big data, requires adequate infrastructure and trained personnel, which can be a challenge for companies that do not have the necessary resources (Cano et al., 2020).

Another major challenge is resistance to change, both at the organizational and individual level. Digitalization implies a transformation in business culture and in the way employees perform their tasks, which can generate uncertainty and resistance in the workforce. To overcome this obstacle, it is essential for companies to implement training programs that enable employees to acquire the necessary skills to adapt to the new digital reality. In addition, it is necessary to foster a culture of openness to change and collaboration, in which employees see digitalization as an opportunity for growth and professional development (García & Fernández, 2020).

Finally, digitization poses challenges in terms of privacy and information security. The collection and analysis of large volumes of personal and business data represent a significant risk if adequate data protection measures are not implemented. Companies must comply with privacy regulations, such as the General Data Protection Regulation (GDPR) in Europe, and must invest in cybersecurity systems that ensure information protection. Customer trust is a key factor in the success of digital business models, and any privacy breach can have a negative impact on the company's reputation and sustainability (Sanchez & Lopez, 2019).

5.2 Ethical and privacy challenges in the digital age

Digitization and the use of advanced technologies such as artificial intelligence (AI) have generated enormous benefits in terms of efficiency and access to information, but have also raised serious ethical concerns. These ethical challenges focus primarily on data privacy, algorithmic bias and information security, three critical areas that organizations must manage to ensure responsible use of technology. In an environment where the use of data is driving innovation, the challenge lies in balancing the opportunities of digitization with the protection of users' fundamental rights and values.

Data privacy: a right at risk

Data privacy is one of the most hotly debated ethical issues in the digital age. Organizations collect and analyze large volumes of personal information, ranging from behavioral patterns to biometric data, to improve their services, optimize their performance and deliver personalized experiences. However, the massive collection of data poses the risk of invading individuals' privacy and using this information unethically. In many cases, users are not fully aware of how much data is collected about them and how it is used (Gonzalez & Fernandez, 2019).

The General Data Protection Regulation (GDPR), adopted in the European Union in 2018, is an effort to regulate the use of personal data and ensure that organizations respect users' privacy rights. However, compliance with these regulations remains a challenge for many companies, especially in cases where data collection is indispensable to the operation of the business. The lack of transparency and control by users in the handling of their data creates a significant ethical concern, as companies could profit from the exploitation of personal data without ensuring that the rights of individuals are adequately protected (López & Sánchez, 2020).

On the other hand, the use of facial recognition and biometric analysis technologies has also generated ethical controversies. Although these technologies can improve security and efficiency in some contexts, they can also be invasive and pose a threat to individual privacy. For example, the use of facial recognition systems in public spaces allows organizations and governments to monitor citizens without their explicit consent, raising questions about the right to privacy and the risk of mass surveillance

(Pérez & Gómez, 2021).

Algorithmic bias: a challenge for fairness

Algorithmic bias is another major ethical challenge in the digital age. AI algorithms are trained on historical data and behavioral patterns, which implies that they can inherit biases and prejudices present in that data. This algorithmic bias can result in unfair or discriminatory decisions in critical areas such as personnel selection, credit granting, or administration of justice (Rodriguez & Moreno, 2021).

For example, in human resources, AI systems used to select candidates may rely on historical data that reflect discriminatory practices, thereby perpetuating gender or racial bias. Unless there is thorough monitoring and proper tuning of algorithms, algorithmic bias can reinforce inequality and exclusion of certain groups. This ethical concern not only affects workplace equity, but also raises questions about the responsibility of organizations and AI designers to ensure that their systems are fair and inclusive (Garcia & Torres, 2020).

The lack of transparency in AI algorithms also exacerbates the problem of bias, as many of these systems operate as "black boxes" where users do not understand how decisions are made. This makes it difficult to identify and correct bias, and raises the challenge of implementing auditing and oversight practices that ensure fairness and accountability in the use of AI. Transparency and explainability have become essential elements in mitigating algorithmic bias and ensuring that AI systems operate ethically and fairly.

Data security: the threat of cyber-attacks

Data security is another key issue in the digital environment, as organizations handle massive volumes of sensitive information that can be vulnerable to cyber-attacks. Cyber attacks pose a threat to both the privacy of users and the integrity of organizations, and can have devastating consequences in terms of loss of trust, economic damage and violation of rights. Organizations should invest in robust security measures, such as data encryption, network monitoring, and staff training, to protect against potential threats (Hernandez & Rodriguez, 2021).

In some cases, cyber attacks can be used to steal sensitive information or manipulate AI systems for the benefit of the attackers. For example, adversarial attacks, which alter input data to manipulate AI algorithm decisions, can have serious consequences in sectors such as healthcare, banking or national security. These risks require organizations to adopt rigorous cybersecurity policies and establish rapid response protocols to mitigate the impact of attacks.

5.3 The role of leadership in technology integration

In an environment marked by digitization and the adoption of advanced technologies, organizational leadership plays a fundamental role in the integration of technology into the business model. The transition to digitization and the use of AI requires not only an investment in infrastructure and tools, but also a cultural change that can only be achieved under the direction of leaders committed to developing a strategic technology vision. Organizational leaders must guide this transformation with a clear vision and a focus on innovation, ensuring that the implementation of technology is done responsibly and aligned with the objectives and values of the organization (Vargas & Sanchez, 2019).

The strategic vision and the definition of objectives

For effective technology integration, leaders must develop a strategic vision that addresses the short-, medium- and long-term goals of the organization. Implementing technologies such as AI and big data requires careful planning and a thorough understanding of industry-specific opportunities and challenges . Leaders must be able to identify how these technologies can improve processes, optimize performance and deliver value to customers. This vision must be aligned with the company's values and objectives, so that digitization is not simply a technical change, but a strategic transformation (Fernandez & Garcia, 2020).

In addition, leaders should establish clear and measurable goals to assess the impact of technology on organizational performance. Defining specific performance indicators allows the organization to monitor the success of technology integration and make adjustments if necessary. Leaders play a crucial role in creating a data-driven culture, where decisions are based on objective information and rigorous analysis to guide the company's direction in a changing environment.

Communication and change management

One of the main challenges in technology integration is change management, and in this regard, the role of leadership is essential. Digitalization implies significant changes in the way employees perform their work, and some are likely to feel insecure or resistant to these transformations. To overcome these barriers, leaders must promote transparent and constant communication, explaining to employees the reasons behind the implementation of new technologies and how these changes will benefit the organization and their individual careers (López & Fernández, 2021).

Change management requires leaders to take an inclusive approach, involving employees in the transformation process and encouraging them to voice their concerns and suggestions. Creating an environment where workers feel part of the change is critical to reducing resistance and ensuring a smooth transition. Leaders must act as facilitators, providing the necessary support and offering training opportunities for employees to acquire the skills required in the digital environment.

Training and competency development

The integration of technology in organizations requires employees to develop new technical skills, such as how to manage AI systems and data analytics. In this regard, leaders have a responsibility to ensure that employees have the resources and training necessary to adapt to new technological demands. This includes investing in training and professional development programs that enable workers to improve their skills and acquire specialized knowledge in the use of digital tools.

Continuous training is also essential to maintain the organization's competitiveness in an ever-changing environment. Leaders must foster a culture of continuous learning, where skills development is valued as a long-term strategic investment. This approach not only benefits the organization, but also increases employee satisfaction and commitment by providing employees with the opportunity to grow and adapt in a dynamic work environment (Hernández & Gómez, 2021).

Ethical leadership and accountability in the use of technology

Finally, leadership in the digital age also implies an ethical responsibility in the use of technology. Leaders must ensure that the implementation of AI and other digital tools respects the organization's core values and promotes transparency, fairness and data privacy. This requires establishing policies for responsible use of technology, including measures to avoid algorithmic bias, ensure data security, and protect user privacy (Rodriguez & Morales, 2020).

Ethical leadership also involves being aware of the social impacts of digitalization, especially in terms of employment and inequality. Leaders must anticipate the effects of automation and take steps to minimize job displacement, promoting the relocation and training of affected employees. This responsible approach enables the organization to adapt to technology in an ethical and sustainable manner, building an organizational culture that values both innovation and the well-being of its employees and customers.

5.4 Future scenarios for human-machine collaboration

Human-machine collaboration is becoming a central feature of the work environment and is expected to continue to evolve as advanced technologies such as artificial intelligence (AI), robotics and automation become further integrated into business processes. This interaction, which goes far beyond basic automation, is characterized by a symbiotic relationship in which both humans and machines contribute their strengths to achieve common goals. Future scenarios for human-machine collaboration present themselves as a field of opportunities, but also as a terrain full of ethical, operational and social challenges.

Future scenarios in human-machine collaboration focus on improving efficiency, productivity and creativity in the workplace. To achieve these goals, it will be crucial to strike the right balance in the use of technology, ensuring that AI intervention and automation do not displace the human role, but rather enhance it. This chapter explores four key scenarios that reflect the opportunities and challenges of human-machine collaboration: synergistic work, augmented automation, creative innovation, and sustainability.

Scenario 1: working in synergy

The first future scenario for human-machine collaboration is based on the concept of synergistic work, where humans and machines work together on processes that require both human intelligence and the analytical capabilities of AI. In this context, machines do not replace human workers, but complement them. Repetitive and analytical tasks are handled by algorithms and robots, allowing human employees to focus on activities that require creativity, judgment and empathy.

This approach is already seen in industries such as healthcare and education, where machines assist professionals with diagnostic tasks and data analysis, while humans provide the empathy and judgment needed for decision making (Davenport & Kirby, 2016). For example, AI systems can analyze large amounts of medical data and detect patterns that might go unnoticed by clinicians. However, the final decision and treatment of the patient still depends on the expertise and judgment of the healthcare professional.

In the future, this type of synergistic collaboration will be key in virtually every industry. Organizations will need to find ways to integrate technology so that human employees can work hand-in-hand with automated systems, allowing each party to bring the best of their capabilities to the table. For this model to work, it will be essential for workers to have digital and technology skills that enable them to collaborate effectively with AI and other advanced technologies. Continuous training will therefore be a central element in this scenario of working in synergy (Autor et al., 2020).

Scenario 2: increased automation

Augmented automation is another scenario in which a future of human-machine collaboration is on the horizon. This concept refers to the idea that automation does not simply replace human functions, but rather augments the worker's capabilities by providing tools that enhance their skills and improve their performance. Instead of seeing automation as a tool that replaces the human being, in this scenario it is seen as a resource that allows him to perform his tasks faster and more efficiently.

In sectors such as advanced manufacturing and logistics, Collaborative robots (cobots) and AI systems are already being used to augment human capabilities in physical and operational tasks. Cobots enable workers to operate in safe environments, while improving the accuracy and speed of tasks. For example, on a production line, cobots can handle heavy material handling or precision tasks, while the worker oversees and coordinates the overall process.

Augmented automation is also applicable in areas of intellectual work, such as data analysis and decision making. AI systems can process large amounts of information in seconds, providing recommendations and predictions that facilitate strategic decision-making. However, the final judgment and strategy remain in the hands of humans, who interpret the results and consider intangible factors that machines cannot evaluate, such as the social and cultural context of decisions (Brynjolfsson & McAfee, 2014).

Scenario 3: creative innovation

One of the most fascinating aspects of human-machine collaboration is the potential to drive creative innovation. Although creativity has traditionally been considered an exclusively human capability, the development of AI algorithms that can learn patterns and generate innovative ideas has opened up new opportunities in the field of creativity. In this scenario, AI acts as a catalyst for innovation, providing suggestions, exploring new possibilities and offering insights that can inspire humans.

In the design and advertising industry, for example, AI tools are already being used to analyze consumer trends and generate creative ideas based on user behavior patterns. Companies can create personalized and optimized advertising campaigns by analyzing data in real time, allowing them to tailor their messages to consumers' needs and desires. However, human creativity remains indispensable to give an authentic and emotional touch to the final message, an element that machines cannot yet replicate (González & López, 2019).

The future of collaboration in creative innovation will be characterized by the development of tools that allow humans to interact with AI in a collaborative and dynamic way. AI systems can generate proposals and prototypes based on data, while humans interpret these results and adapt them to specific cultural and social contexts. Creativity will be a collaboration in which humans and machines bring their intuition, analysis and experience to create original and relevant solutions.

Scenario 4: collaboration for sustainability

Another key future scenario in human-machine collaboration is the use of technology to promote sustainability. As awareness grows about the effects of climate change and the need for sustainable development, organizations are looking for ways to reduce their environmental impact and operate more responsibly. In this context, AI and automation can play a key role in efficient resource management, process optimization and waste reduction.

For example, in the energy sector, AI makes it possible to monitor resource consumption in real time and adjust production according to demand, reducing energy waste. In the agriculture sector, the technology helps to optimize the use of water and fertilizers, reducing environmental impact and increasing efficiency in the use of natural resources (Rodríguez & Morales, 2021). AI systems can also predict weather events and changes in environmental conditions, allowing companies to adjust their operations to minimize negative effects on the environment.

Collaboration for sustainability also implies a change in the way organizations value their operations and their impact on society. In the future, employees are expected to work alongside machines on projects that prioritize the well-being of the planet, using technological tools to measure and reduce ecological impact. This scenario presents an opportunity for meaningful human-machine collaboration that contributes to sustainable development and environmental stewardship.

Challenges in human-machine collaboration scenarios

Although future scenarios for human-machine collaboration offer numerous opportunities, they also pose significant challenges that must be addressed. One of the main obstacles is resistance to change on the part of employees, who may feel that their jobs are threatened by automation. For collaboration to be successful, it is critical that organizations implement training and retraining policies that help employees adapt to the new roles demanded by technology.

Another challenge is the need to ensure fairness and transparency in AI systems used in the work environment. Algorithmic bias and lack of explainability of some algorithms can affect decision making and create issues of fairness and transparency in the workplace. Organizations will need to develop ethical and governance frameworks that ensure that AI is used in a fair and responsible manner, minimizing the risk of discrimination and promoting equal opportunities (López & Fernández, 2020).

Finally, moving towards these future scenarios will require a robust digital infrastructure and adequate cybersecurity measures. Human-machine collaboration involves a constant exchange of information between humans and digital systems, which increases vulnerability to cyberattacks and security threats. Companies must invest in technology infrastructure and cybersecurity strategies to protect both employees' personal data and the integrity of AI systems.

Human-machine collaboration represents a fundamental transformation in the world of work and presents future scenarios full of possibilities and challenges. The four key scenarios-synergistic work, augmented automation, creative innovation, and collaboration for sustainability-illustrate how organizations can leverage technology to improve efficiency, creativity, and environmental responsibility. However, successful implementation of these scenarios will require a strategic and ethical approach by organizations, which will need to foster empowerment, transparency and privacy protection in their operations.

Human-machine collaboration offers an unprecedented opportunity to reimagine work and achieve significant breakthroughs across industries. Organizations that responsibly and ethically embrace these approaches will be well positioned to thrive in an increasingly digitized work environment, where humans and machines work together to create a more efficient, just and sustainable future of work.

References

Acemoglu, D. and Restrepo, P (2018). Artificial intelligence, automation, and work. Journal of Economic Perspectives, 33(2), 193-210.

Acemoglu, D. and Restrepo, P. (2020). Automation and new tasks: How technology displaces and reinstates work. Journal of Economic Perspectives, 33(2), 3-30.

Acemoglu, D. and Restrepo, P. (2020b). The race between man and machine: Implications of technology for growth, factor shares, and employment. AmericanEconomicReview, 108(6), 1488-1542. https://doi.org/10.1257/aer.20160696

Accenture (2020). Harnessing AI responsibly: The ethical and social impacts of artificial intelligence. Retrieved from https://www.accenture.com

Author, D. (2015). why are there still so many jobs? The history and future of automation in the workplace. Journal of Economic Perspectives, 29(3), 3-30. https://doi.org/10.1257/jep.29.3.3

Autor, D. H., Levy, F., & Murnane, R. J. (2003). The skill content of recent technological change: An empirical exploration. Quarterly Journal of Economics, 118(4), 1279-1333.

Autor, D. and Dorn, D. (2013). The growth of low-skill service jobs and labor market polarization in the United States. American Economic Review, 103(5), 1553-1597. https://doi.org/10.1257/aer.103.5.1553

Bass, L., Weber, I., & Zhu, L. (2015). DevOps: A perspective for software architects. Addison-Wesley.

Bessen, J. (2019). AI and jobs: The role of demand. NBER Working Paper No. 24235. https://doi.org/10.3386/w24235.

Brynjolfsson, E. and McAfee, A. (2014). The second machine age: Work, progress, and prosperity in an age of shining technologies. W.W. Norton & Company.

Brynjolfsson, E. and McAfee, A. (2017). Machine, platform, crowd: Leveraging our digital future. W.W. Norton & Company.

Cano, F., González, A. and López, M. (2020). Artificial intelligence and productivity in manufacturing industry. Journal of Industrial Engineering, 15(2), 45-62.

Cardenas, A. A., Amin, S., & Sastry, S. (2018). Research challenges for control system safety. National Institute of Standards and Technology.

Deloitte (2021). Global human capital trends report 2021. Deloitte Insights.

Deming, D. J. (2017). The growing importance of soft skills in the labor market. The Quarterly Journal of Economics, 132(4), 1593-1640.
https://doi.org/10.1093/qje/qjx022

Fernández, A. and Martínez, L. (2021). Ethics and privacy in artificial intelligence. Journal of Technological Innovation, 15(2), 123-136.

Ford, M. (2015). The rise of the robots: Technology and the threat of a jobless future. Basic Books.

World Economic Forum (2018). Future of jobs report 2018. Retrieved from https://www.weforum.org

World Economic Forum (2020). Future of Jobs 2020 report. Retrieved from https://www.weforum.org

Garcia, E. (2020). Automation in the Industrial Revolution. Journal of Industrial Engineering, 8(3), 112-129.

García, J. and Rodríguez, M. (2021). Work transformation and ethics in the age of artificial intelligence. Journal of Applied Ethics, 12(4), 67-80.

García, M. and Fernández, P. (2020). Digital transformation and sustainability: Artificial intelligence as a key tool. Revista de Administración, 8(1), 88-103.

González, J. and Ortiz, R. (2019). Strategies for personalizing the customer experience using AI. Journal of Business and Technology, 12(3), 71-86.

Herrera, C. (2019). Efficiency and ethics in the use of intelligent machines. Ethics and

Technology, 5(3), 201-218.

López, A. (2019). The revolution of the machines. Editorial Tecnología Industrial.

López, R. (2022). Algorithmic discrimination and its social implications. Journal of Social Studies, 10(3), 157-170.

Manyika, J., et al. (2017). A future that works: Automation, employment and productivity. McKinsey Global Institute.

Martinez, R. (2018). Impact of digital technology on the workplace. Journal of Technology and Work, 15(2), 45-59.

McAfee, A. and Brynjolfsson, E. (2017). The second machine age: Work, progress, and prosperity in an age of brilliant technologies. W.W. Norton & Company.

McKinsey & Company (2018). Changing skills: Automation and the future of the workforce. Retrieved from https://www.mckinsey.com

McKinsey & Company (2021). Achieving equity in AI systems: Addressing bias and transparency. Retrieved from https://www.mckinsey.com

Perez, D. (2019). Responsibility and accountability in the age of artificial intelligence. Law and Technology, 7(4), 301-320.

Pérez, J. (2020). AI and robotics: The new era of work. Editorial Avances Tecnológicos.

Perez, L. and Martin, C. (2019). Effects of automation on productivity and cost reduction in manufacturing industry. Economic Studies, 28(4), 27-49.

PwC. (2021). Global artificial intelligence study: Harnessing the AI revolution. Retrieved from https://www.pwc.com

Russell, S. and Norvig, P (2021). Artificial intelligence: A modern approach (4.ª ed.). Pearson.

Sanchez, P and Lopez, E. (2019). Artificial intelligence in the pharmaceutical industry: A focus on quality control. Health Sciences Research, 3(1), 14-29.

Topol, E. J. (2019). Deep medicine: How artificial intelligence can make health care human again. Basic Books.

van der Aalst, W. M. P (2016). Process mining: data science in action (2nd ed.). Springer. https://doi.org/10.1007/978-3-662-49851-4

Vera, S. (2021). Robots and efficiency in the automotive industry. Technological Analysis, 19(1), 52-68.

Willcocks, L. P., Lacity, M. C., & Craig, A. (2015). Robotic process automation: The next transformational lever for shared services. Journal of Information Technology Teaching Cases, 5(2), 77-87. https://doi.org/10.1057/jittc.2015.5

Printed by Books on Demand GmbH, Norderstedt / Germany